Restoring the Natural Order: A Scientific Assessment of Floodplain Reconnection

Faris

Table of Contents

Chapter 1: Introduction

Stream restoration techniques have been developed and implemented since the early 19th century (Wohl et al., 2015). Originally, restoration was used to improve the aesthetic of a waterway or to expand recreation; this ultimately resulted in an aquatic system that was uniform, physically simplified, and ecologically less diverse (Poff et al., 2007). Ecological engineers began to rethink their techniques and shifted efforts towards restoration practices that improved habitat for aquatic life rather than for human use (Gowan & Fausch, 1996). Since laws were created to protect waterways from pollution in the United States and in other countries, water quality became a focal point of restoration. Water quality improvement tends to focus on mitigation of point and non-point source pollution, improvements to riparian corridors and floodplains, changes to discharge rates, or altered channel morphology (Wohl et al., 2015).

Recently, an emphasis on improving stream function over form has taken precedence. An emergent form of restoration known as floodplain reconnection prioritizes channel-floodplain connectivity to improve longitudinal linkage of the waterbody, sediment and nutrient fluxes, and ecological productivity (Palmer et al., 2010). This restoration method is in its infancy relative to others, so it is important to build on the current dataset for the purpose of understanding its ecological impact. The objective of this study was to characterize the impact of the floodplain reconnection method on streams in southwestern Pennsylvania by analyzing hydrology, sediment dynamics, and nutrient cycling. These data were compared to data from unrestored sites in the same region to characterize the effect of floodplain reconnection as a method of restoration.

1.1 Natural Channel Design

The perspective that restoration should focus on improving the physical characteristics of a stream is based on an assumption that once a channel is adapted to new flow rates and sediment fluxes post-restoration, ecological processes (e.g., biodiversity, nutrient cycling, food web connectivity) will be restored (Palmer et al., 2014). Dave Rosgen, a prominent scholar of river restoration, is known as a pioneer of this type of design method. He developed the Natural Channel Design (NCD) template, an outline that suggests a physical channel design for a stream without considering what is best for aquatic stability. This method of restoration has been criticized by the scientific community because channel classifications tend to be arbitrary and do not consider the dynamic nature of a waterbody.

This method also fails to address the idea that the existing condition of a channel likely does not reflect the original, or the optimal condition for the future (Kondolf, 1995). The NCD operates on the basis that stream dimension, pattern, and profile need to be restored to match the physical properties of a natural and stable stream (Rosgen, 2007). Rosgen believed that this method would restore the chemical, biological, and physical aspects of a stream because it will exhibit a stable channel design. However, the NCD does not address chemical or biological processes that are imperative for optimizing stream function (Palmer et al., 2014).

Scientists are rallying for process-based approaches to channel design over the form-based practice that is associated with the NCD (Simon et al., 2007). A study by Walter and Merritts (2008) indicated that NCD will not encourage the return of previously established species nor the ecological services that they previously provided. The authors also suggest that modern stream and wetland restoration should focus on ecological design principles that aim to restore ecosystem functions and services (Walter & Merritts, 2008). In a comprehensive examination of the problems associated with NCD, some restoration scientists claim that form-based restoration methods ignore the necessity of open aquatic systems that can adjust to altered inputs of energy and material inputs (Simon et al., 2007). While NCD can still be a useful restoration model, it needs to be considered in conjunction with ecosystem-wide processes to be effective.

1.2 Restoration Through Floodplain Reconnection

To address concerns about form-based restoration, the idea of floodplain reconnection should be considered. Restoration scientists and engineers have developed an approach that rejoins the stream channel with its adjacent floodplain (Harrison et al., 2014). The goal of floodplain reconnection is to reduce the difference between base and peak flow in the channel, something that can limit erosion and increase retention time to encourage physical and biological processes (McMillan & Noe, 2017). This restoration method often focuses on using off-channel water retention areas (like wetland floodplains) to encourage the absorbance and settling of contaminants that are attached to fine sediment (Wohl et al., 2015). In connecting the channel to the floodplain, the floodplain receives water during periods of high flow, something that happens more readily than it would in pre-restoration conditions. This reconnection can improve biodiversity in the ecosystem due to the reprieve that the slow moving, vegetated, non-turbid floodplain waters provide for stream biota during flooding events. Research from Bayley (1991) and Welcomme (1979) showed that this increase in biodiversity associated with the presence of floodplains leads to higher fish yields per area, making rivers with floodplains some of the largest freshwater fisheries.

Channel designs can vary based on available space; these designs might include several levels of flat floodplains with transitional steps (inset floodplain method) or can be comprised of a shallow bank slope that connects the channel with the bank (McMillan & Noe, 2017) (Figure 1). In urban areas where the inset floodplain method may be used, the floodplains will generally have dense vegetation and high groundwater tables. These characteristics provide carbon and reduce the potential of permanent nitrogen removal via denitrification (Kaushal et al., 2008). The presence of dense vegetation also enhances sediment and particulate phosphorus retention (Davis et al., 2015).

Figure 1

Urban Versus Rural Floodplain Reconnection

Note: A stream constructed for the purpose of floodplain reconnection might look different depending on the setting. In an urban environment, multiple floodplains may be constructed. When there is more room for construction, a wide floodplain may be built. This figure was redrawn based on a graphic in McMillan and Noe (2017).

While floodplain reconnection restoration can improve the ecological function of waterways, the amount of success achieved can be limited if land-use stressors within the watershed are ignored during planning and construction (Wohl et al., 2015). A study by Roley et al. (2012) illustrates this phenomenon with an example in which an agricultural stream in Indiana with an inset floodplain near a channelized ditch exhibited only minor improvements in nutrient regulation over time because excess nitrate pollution from croplands within the watershed remained a problem post-restoration. It is important to consider ecosystem stressors at a watershed scale to ensure that the extent of restoration will be successful. The following sections provide a review of the characteristics of floodplain reconnection pertaining to hydrology, sediment dynamics, and nutrient cycling.

1.2.1 Hydrologic Connectivity

Hydrologic connectivity describes the surface and subsurface flow of water in an aquatic system. Understanding how water flows is important for conceptualizing water storage. Restoration through floodplain reconnection includes adding woody debris to the stream so that water storage can be improved. Rana et al. (2017) use salt tracer injections to quantify the effect that in-stream debris has on solute storage as a benchmark for water storage. The authors demonstrated that in-stream debris helped increase transient storage zones, but variations in flow rate would further alter the rate of transient storage (Rana et al., 2017). Improvements in water storage cause increased residence time of water, something that facilitates biological processes (Hester et al., 2013).

Another hydrologic goal of floodplain reconnection is to slow water velocity. Floodplain reconnection aims to build an ecosystem that is resilient to problems associated with flooding, such as increased erosion and water velocity, both of which can lead to excessive sediment and nutrient loading (McMillan & Noe, 2017). This relationship can be further understood by considering the flood pulse concept, which is described as, "A spectrum of geomorphological and hydrological conditions…" that impacts the "…existence, productivity, and interactions of the major biota in river-floodplain systems" (Junk et al., 1989, p. 110). This idea is important to consider when thinking about the impacts that floods have on water velocity in aquatic ecosystems;

restoration options such as the NCD do not necessarily account for ecological issues that spur from high peak flows.

Prioritizing paths of hydrologic connectivity during restoration is pivotal to the ecological success of the project. With variable amounts of precipitation, channels often have changing hydrologic connectivity, which means that multiple geomorphic zones can exist laterally across floodplains. These zones function to control moisture, organic matter content, and vegetation, all of which influence rates of biogeochemical processes like nutrient deposition, mineralization, and denitrification (Noe et al., 2013). This emphasizes the point that changes in water depth along with base and peak flow rates can have a multitude of effects on an aquatic system. Considering the effect of floods ensures that restoration efforts will take into account the complex interactions that changes in hydrology have on sediment and nutrient dynamics.

1.2.2 Sediment Dynamics

The flood pulse concept states that hydrology impacts sediment transfer in aquatic systems because the velocity and volume of water can move sediment at different rates (Junk et al., 1989). In areas of high water velocity, coarse-grained material is likely to be deposited where the opposite is true in areas of low velocities (Hupp et al., 2013). The Mid-America Regional Council (MARC) suggests that understanding the movement of sediment in aquatic ecosystems is important for several reasons. Excessive amounts of sediment can increase flooding potential in certain areas of the channel, it can lead to turbid water which disrupts the photic zone and prevents vegetation from growing, and it disrupts food chains by destroying habitat for macroinvertebrates, something that causes declines in fish population (Mid-American Regional Council [MARC], n.d.). Sediment can also harbor and transfer excessive amounts of nutrients like ammonium, organic nitrogen, and phosphorus, a phenomenon that is included in the following overview of nutrient transfer (Davis et al., 2015).

There are several ways in which floodplain reconnection limits excessive sediment transfer, and one of them is by improving riparian zones. Floodplain vegetation found within the riparian zone acts as a barrier that slows water velocity and aids in sediment deposition outside of the channel (Davis et al., 2015). The physical characteristics of channels are also important to consider when identifying modes of sediment transfer. McMillan and Noe (2017) found that sedimentation rates were positively correlated with the bank height to floodplain width ratio. They explain this relationship as a function of sediment supply in the channel, which is associated with physical condition and location within the greater river network. The sites that exhibited the highest rates of sedimentation in their study were located downstream of heavily incised stream sections identified by significant bank erosion (McMillan & Noe, 2017).

Legacy sediment is another aspect of sedimentation that can be remedied by floodplain reconnection. This term refers to the presence of sediment in an ecosystem that is associated with historical human alteration of an area. Some restoration projects work to remove this sediment to reconnect channels with floodplains in an effort to truly restore the waterbody to its natural state before human intervention (Wohl et al., 2015). Many streams in the Mid-Atlantic Region of the United States have been contaminated with legacy sediment from milldams, structures that promote sediment deposition upstream and scouring downstream. A study on Little Conestoga Creek in Pennsylvania measured sedimentation rates, bank erosion rates, and channel morphology and compared these data to the proximity of the site to a milldam. The researchers found that the majority of the sediment deposited via erosion came from sites nearest the milldam (Schenk & Hupp, 2009).

A restoration project within the Big Spring Run watershed removed legacy sediment from the stream banks. Hartranft (2019) found that by restoring this area through the floodplain reconnection method, erosion was decreased, sediment deposition was limited to a few centimeters, and water quality improved. This restoration method was observed to be 10.5 times more effective at reducing suspended sediment loads than other agricultural best management practices (Hartranft, 2019). After a few years, the improved environment supported biology through several metrics. Salamander species and green frogs inhabited the site, and pollution-intolerant fish species like the rosy side dace moved back into the area. Diatom species diversity had also increased; when the study was published, there were fewer species associated with pollution tolerance than before restoration (Hartranft, 2019).

The issue of sedimentation in sites that have been impacted by humans is further emphasized when examining sites in this region that have not been altered by human activity. In the Big Spring Run watershed site, archaeological, historic, and geomorphological evidence showed that unimpaired, pre-settlement valley bottoms in the region were wetlands. Walter et al. (2013) studied valley bottoms that were never impacted by damming or substantial sedimentation from agriculture and came to the conclusion that restoration efforts in this area should focus on converting floodplains to wetlands. Understanding the presence of legacy sediment allows for restoration scientists to account for the existence of human intervention in a system, something that will clarify the natural history of the ecosystem.

1.2.3 Nutrient Cycling

Nutrient retention is a goal of floodplain reconnection restoration. Tracking nutrient transfer in aquatic restoration projects is important because nutrients such as phosphorus and nitrogen are necessary for ecosystem health, but too much or too little can result in the loss of biodiversity in a waterbody. Nutrient transfer can be complex; rates of nutrient retention, removal, and release are controlled by many different biotic and hydrologic factors. McMillan and Noe (2017) noted that nutrient processing rates in floodplain ecosystems were primarily controlled by gradients in nutrient supply from pools within the soil, nutrient inputs from outside of the floodplain, geomorphic features of the site, and age of the restoration project. This implies that aquatic ecological productivity emerges in streams that are connected to their floodplains. The researchers found that inset floodplains trap large amounts of nutrients, something that reduces downstream exports (McMillan & Noe, 2017). Nutrient removal in general is highly influenced by its connection with stream water and storm water runoff. Along with biogeochemical processing rates, nutrient loading and removal can be measured as a function of this connectivity (Noe et al., 2013). This section is divided into characteristics of nitrogen and phosphorus cycling that should be considered before commencing a restoration project.

Nitrogen Transfer. Nitrogen (N) can enter streams from runoff, groundwater, and atmospheric deposition; the extent that nitrogen is delivered downstream depends on the permanent removal and temporary storage of N. Permanent removal of N occurs during denitrification, or the reduction of nitrate to gaseous forms (Craig et al., 2008). In aquatic systems with high water tables and extended time of inundation, reducing conditions that favor denitrification are often created (Roley et al., 2012). In unrestored headwater urban floodplains, short residence times and lowered groundwater tables can conversely limit redox dependent processes. Temporary storage of N can happen through biological and physical methods. Biota can store N until decomposition occurs. Physical storage can happen at areas of inundation where water is not flowing. In these cases, N can be retained and later returned to surface waters (Craig et al., 2008).

A study by Hefting et al. (2003) on soil nitrogen cycling in riparian wetlands details mechanisms of nitrogen transfer. Riparian zones are productive ecosystems that cycle N at a high rate. Groundwater table fluctuation regulates processes in the N cycle due to its influence on aerobic and anaerobic conditions in the soil. The researchers compared the effects of these processes on nitrogen cycling in European riparian zones and found that water table elevation is the leading determinant of N cycling and its end products. Hefting et al. (2003) report that when the water table level is within 10 cm of the soil surface, ammonification is the leading process of N cycling. When the water table is between 10 and 30 cm from the soil surface, denitrification is favored and N availability is reduced. When sites are dry and the water table is more than 30 cm from the soil surface, high net nitrification leads to the accumulation of nitrate (Hefting et al.,

2003). This research exemplifies the importance of inundated wetland zones for regulating mechanisms of N cycling.

Craig et al. (2008) propose a framework for choosing sites and approaches to ensure that the best possible benefits of nitrogen removal can be met in areas where it is needed. The researchers note that site selection should include consideration for watershed location, history, and landscape characteristics along with feasibility of the project. The study outlines the major modes of transportation for N as the following: runoff, interflow, groundwater, and artificial drainage systems. Flow paths depend on land use, geology, aquifer geometry, hydraulic gradients, soils, and drainage infrastructure. The authors also underscore the importance of tailoring nitrogen removal to the size of the stream; nitrogen uptake rates tend to increase as stream order and discharge decrease. Identifying discharge rates that deliver N is also paramount. Restoration scientists should identify the flow at which most N is being delivered to create a design that will facilitate N removal in environmental where it is needed (Craig et al., 2008).

Phosphorus Transfer. Phosphorus (P) is a key limiting nutrient in freshwater ecosystems that is transported from uplands by subsurface and surface flow (Domagalski & Johnson, 2012). Because of this, the ability of wetlands and streams to retain P is important when considering downstream water quality. Like N, P retention is impacted by biological, physical, and chemical properties. Wetlands, floodplains, and associated streams can store P and also transform it into biologically available forms (and vice versa). In wetlands, floating macrophytes can absorb P from the water column and store it in their tissues (Mitsch et al., 1995). When the vegetation decomposes, that P is released back into the water. Other types of wetland plants store P in their roots and rhizomes; these plants uptake P from soil porewater rather than the water column. However, P is primarily transported downstream by the flow of water. When a system is flooded, water interacts with the floodplain and in the presence of a wetland, P can be transported in and out of the floodplain at greater rates. P can also bind to sediment and be deposited in the floodplain wetland during floods which enhances P storage (Reddy et al., 1999).

The mobility of P was studied by Surridge et al. (2012) who analyzed data from a range of hydrological events in a long-restored floodplain site. They found that after floodplain inundation, P was released from sediments and was taken up by the pore water and surface water; the concentration of P in porewater was higher than in surface water, and both were higher than rainwater or floodwater adjacent to the river. They also found that short-term hydrological events led to increases in dissolved P concentrations within the floodplain. This research suggests that restored floodplains could be sources of P to adjacent waterbodies (Surridge et al., 2012). This specific case illustrates the impact that P transfer has on floodplain restoration areas.

1.3 Summary and Project Goals

This background information identified ecologically important aspects of stream restoration. Historical methods of restoration, like the NCD, have proven to be less effective than anticipated for processes like water, sediment, and nutrient storage. Since then, restoration scientists are moving away from tactics that focus on restoring the physical form of a stream and towards those that work to improve ecological function of the entire ecosystem. Floodplain reconnection aims to create a system that is resilient to varying hydrologic conditions. Aquatic ecosystems should be restored to a pre-disruption state that encourages floodwater retention and appropriate levels of sediment and nutrient retention that will foster healthy biotic communities. Restoration has evolved through time, and it is important that research is continued to achieve the most comprehensive methods for healthy and sustainable freshwater ecosystems. The body of literature on floodplain reconnection restored sites is relatively young. In studying restored streams in the Mid-Atlantic Region of the USA with varying ages of restoration, more data can be collected to better understand the effect of restoration on water storage, sediment dynamics, and nutrient cycling.

This study investigated whether or not enhancing connectivity between the channel and the floodplain improved the ecosystem's ability to act as a sink for water, sediment, and nitrogen and phosphorus. To assess the presence of these characteristics, the following hypotheses were tested in this study:

- We hypothesized that restored sites would have increased water storage and less variable ranges in water depth following precipitation.
- We would also observe reduced sediment flux into the channel as well as lower rates of sediment transport within the channel at restored sites.
- It was also expected that sediment deposition and storage with bound phosphorus would increase in restored sites compared to those that are unrestored.

- Further, we hypothesized that nitrogen and phosphorus flux out of the restored reaches would be decreased, and the denitrification rate would be increased at restored sites.

These hypotheses were tested by analyzing a dataset comprised of flow data, sediment grain size distribution data, and sediment, surface water, and pore water nutrient data at restored and unrestored streams.

Chapter 2: Timeline and Methods

2.1 Study Sites

The research sites spanned three stream size classes: primary headwaters, headwaters, and wadeable (Table 1). Within each size class, three reaches were analyzed—two at restored sites (Figure 2) and one at an unrestored site (Figure 3), all of which are located in southwest Pennsylvania. The unrestored sites are within Ryerson Station State Park, a 1,164-acre park operated by the Pennsylvania Department of Conservation and Natural Resources (PA DCNR, n.d.). These three sites are 100% forested (U.S. Geological Survey, 2016). The six restored sites—previously impacted by mining and channel incision—are located within the nearby Robinson Fork stream system. Restoration began in 2014 and ended in 2016 at these sites. The land adjacent to the channels is comprised of a mix of agricultural and forested land. The drivers for mitigation of these sites include local oil and gas development. The upland areas were historically used for agriculture while the valley slopes were logged, all of which can contribute to unstable banks and excess nutrient deposition that can negatively impact the aquatic ecosystem.

Table 1

Study Sites

Site Name	Size Class	Drainage area	Slope	Aspect	Restoration status	Restoration date
Unit 4D	Primary Headwaters	0.06 sq mi	6.7°	E	Post	2014
McNay	Primary Headwaters	0.08 sq mi	10.8°	S	Pre	
McCully	Primary Headwaters	0.12 sq mi	9.8°	E	Post	2015
Molinari Tributary	Headwaters	0.83 sq mi	14.2°	S	Post	2014-2016
Kent	Headwaters	2.66 sq mi	12.7°	S	Pre	
Beham	Headwaters	3.02 sq mi	12.8°	S	Post	2016
Molinari	Wadeable	14.2 sq mi	12.6°	W	Post	2016
Lebanik	Wadeable	20.9 sq mi	12.9°	W	Post	2016
Dunkard Fork	Wadeable	24.3 sq mi	15.2°	W	Pre	

Note: Study sites categorized by size class, drainage area, slope, aspect, restoration status, and restoration date.

Figure 2

Map of Post-restoration Sites

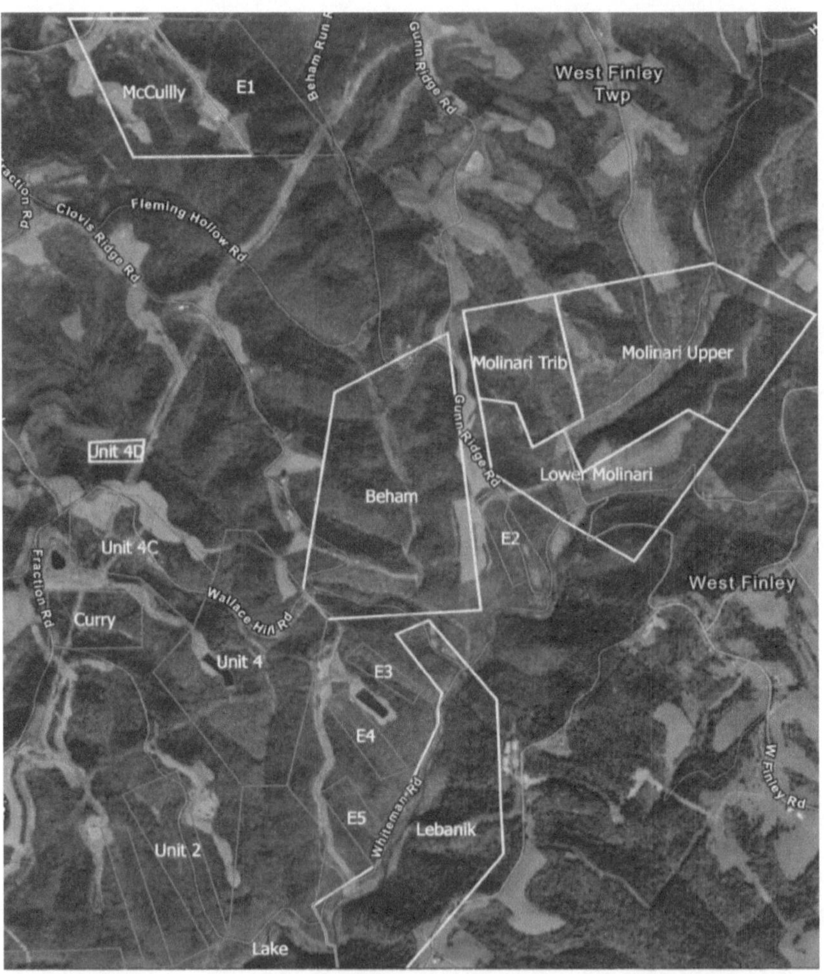

Note: The 6 post-restoration study sites are outlined in yellow. "Molinari Upper" and "Molinari Lower" are outlined separately, but they are considered one site.

Figure 3

Map of Pre-restoration Sites

Note: The three pre-restoration study sites are in Ryerson Station State Park.

2.2 Sampling Timeline

Sampling events were conducted quarterly for the purpose of collecting samples under varying hydrologic conditions. Sampling occurred in July 2020 at low flow, November 2020 during base flow, and March 2021 under high flow conditions. The Wheeling Creek USGS gauge at Elm Grove, West Virginia was used as the water depth reference gauge due to its proximity to the sample sites.

2.3 Field and Laboratory Methods

The goals of data collection were to measure water velocity and storage, sediment deposition and flux, and nutrient composition and flux for the purpose of determining systematic differences between restored and unrestored stream reaches.

2.3.1 Water Depth and Stream Flow

Differences in water depth over time were assessed at post-restoration sites based on historical water level data collected at the downstream end. Water depth data was collected from HOBO piezometers deployed by RES. This data was used to create hydrographs that show water depth over time before and after restoration. These hydrographs characterize the change of each site's response to precipitation before and after restoration.

In wadeable sites, channel flow was measured with a SonTek Flow Tracker. At the primary headwater and headwater streams or during low flow conditions, a Baski cutthroat flume or a pygmy meter was used. Slug injection salt tracer tests were performed so that vadose zone flow could be measured according to methodology in Moore (2003). Salt tracer tests were conducted at each site during every sampling event, except when the flow was too low or too high to effectively and safely conduct the test. Therefore, Unit 4D and Kent in November 2020 was not sampled due to a lack of flow, and Dunkard Fork was not sampled in March 2021 because of high water velocity. A salt solution, comprised of NaCl and stream water, was mixed and added to the stream; the salt concentration was dependent upon stream size (Table 2). A YSI conductivity probe was deployed downstream at a distance that was approximately 25 times as long as the reach is wide to ensure mixing. The meter recorded conductivity spikes in the stream that resulted from the passing salt solution. A 60 mL sample of the slug injection solution was bottled, and a 500 mL sample of stream water was also bottled. The two were placed in a cooler and brought back to the lab at Ohio University so that flow calibrations could be conducted. The curve of conductivity recorded by the YSI meter was converted into salt concentrations using the calibration curve. These data allowed calculation of flow rate and conceptualization of the system's residence time, something that is useful when considering how sediment and nutrients move through a waterway. The difference between channel flow and the flow derived from salt tracing data was used as a measure of transient storage.

Table 2

Salt Tracer Slug Injection Solution

Site	Salt Addition (kg)	Stream Water Addition	
		(liter)	(gallon)
Molinari Trib	1	7	1.8
Unit 4 D	2	14	3.7
McCully	2	14	3.7
Beham	2	14	3.7
McNay	2	14	3.7
Kent	4	28	7.4
Molinari	8	56	14.8
Lebanik	8	56	14.8
Dunkard Fork	10	70	18.5

Note: Research sites and the corresponding makeup of NaCl and stream water needed for the slug injection.

To create a calibration curve in the lab, the site-specific bottles were shaken to ensure that the salt was fully dissolved in the solution. The YSI conductivity meter was calibrated again with 1413 µS/cm solution. Ten mL of the salt solution was added to 100 mL of the stream water, and the background conductivity was measured and recorded. Subsequently, 2 mL of the salt solution was added to the sample five times, and the conductivity was recorded after each addition. Then, 10 mL of the salt solution was added three times and the conductivity was recorded after each addition. This dataset was entered into a pre-made calibration sheet that calculated the slope of the calibration curve relating conductivity to salt concentration. Overall flow was calculated by subtracting the background conductivity from the field data, using the calibration curve to convert conductivity to salt concentration, and integrating these data over time. Calculations accounted for time conversions and background value corrections to determine how much time it took for the salt solution to pass through the reach. During the July 2020 sampling event, the YSI meter failed to record conductivity data in the field at Beham and McCully, so flows were not obtained at these sites. In March 2021, the meter failed to record conductivity measurements at McNay, one of the unrestored streams.

2.3.2 Surface Water Chemistry

A Myron Utrameter II datasonde was calibrated before each day of sampling with pH 4, 7, and 10 calibration solutions as well as 1413 µS/cm conductivity calibration solution. The meter was used to collect surface water measurements for pH, oxidation-reduction potential (ORP), temperature (in degrees Celsius), total dissolved solids (TDS), and conductivity. Surface water samples were collected in the thalweg at the upstream and downstream ends of the restored sites (except for Unit 4D, which is consistently dry at the upstream end) and at the downstream ends of the unrestored sites. Samples were collected for the following tests: general chemistry, dissolved general chemistry, total organic carbon, nitrogen, phosphorus, dissolved nitrogen, dissolved phosphorus, metals, and dissolved metals. A denitrification rate was determined by finding the difference between upstream and downstream total nitrogen concentration derived from the lab samples at the restored reaches. Table 3 expands on the logistics of collecting the surface water samples. Collection protocols followed those outlined by Pennsylvania Department of Environmental Protection (PADEP) (2018).

Table 3

Surface Water Chemistry Tests

	# of Bottles	Bottle size	Test	Acid	Acid amount
Filtered	1	125 ml	Dissolved Nitrogen & Phosphorus	Sulfuric (H$_2$SO$_4$)	2 mL
	1	125 ml	Dissolved Metals	Nitric (HNO$_3$)	2 mL
	1	125 ml	Dissolved General Chem.	None	-
Unfiltered	2	40 ml amber	Total Organic Carbon	Sulfuric (H$_2$SO$_4$)	1 mL
	2	500 ml	Gen Chem.	None	-
	1	125 ml	Metals	Nitric (HNO$_3$)	2 mL
	1	125 ml	Nitrogen & Phosphorus	Sulfuric (H$_2$SO$_4$)	2 mL

Note: A list of the surface water chemistry samples collected, separated by filter requirements, bottle size, test, and preservation method.

The general chemistry samples were collected in two 500 mL HDPE bottles, TOC samples were collected in two 40 mL amber glass VOA vials, and the remaining samples were collected in 125 mL HDPE bottles. Samples being analyzed for dissolved chemistry were filtered with a 0.45 μm filter while the other samples remained unfiltered. Samples were preserved and stored in a cooler at < 4° C before they were sent to PADEP Bureau of Laboratories in Harrisburg, Pennsylvania for analysis. The organic portion of the reported nitrogen concentrations was determined by subtracting the sum of ammonia and nitrate/nitrite from the overall nitrogen value. For organic phosphorus, ortho phosphorus was subtracted from the overall phosphorus concentration. This calculation was used for total and dissolved organic portions.

2.3.3 Pore Water Chemistry

At the restored sites, pore water chemistry was assessed by collecting samples from the saturated floodplain soils at 2 cross-sections per site at the downstream end of the sampling reach. The transect lengths varied by site from 20—40 m. At 4 sampling points at each cross-section, permanently installed micro-Rhizon samplers collected water from the saturated soil of the upper soil layer (0-10 cm soil depth) (Figure 4). The Rhizons were purchased from SoilMoisture, a supplier of tools relating to the relationship between soil and water (SoilMoisture, n.d.). The Rhizons were spaced 2—4 m apart. The water was collected from each Rhizon with a syringe, and each Rhizon had its own syringe to avoid contamination. Once 20 mL of the sample was collected from a single Rhizon, the process was repeated with the remaining Rhizons. The samples from all 8 Rhizons were composited into a 200 mL glass jar before the jar was shaken to mix the sample. Glass graduated cylinders were used to measure out the samples according to their respective tests (Table 4). Pore water samples were not collected at the Beham site because gravel-rich substrate did not allow for installation of Rhizon samplers. The pore size of the Rhizons inherently filtered the samples; therefore, they were not filtered additionally in the field and all data are reported as dissolved. The samples were stored in a cooler and sent to the PADEP lab for nutrient analysis, including dissolved nitrogen, dissolved phosphorus, dissolved metals, general chemistry, dissolved organic carbon, and dissolved inorganic carbon. The organic portion of the reported dissolved nitrogen concentrations was determined by subtracting the sum of ammonia and nitrate/nitrite from the dissolved nitrogen value. For organic phosphorus, ortho phosphorus was subtracted from the dissolved phosphorus concentration.

Figure 4

Rhizon Transect

Note: Image of micro-Rhizon sampler transect pictured at Molinari Tributary, a restored headwaters stream. Blue Rhizons can be seen at the base of the red flags. A sediment pit trap is also pictured in the channel.

Table 4

Pore Water Chemistry Tests

# of Bottles	Bottle size	Test	Min. Volume	Preservation	Preservation amount
1	125 mL	General Chemistry	40 mL	None	-
1	125 mL	Dissolved Nitrogen & Phosphorus	30 mL	Sulfuric (H_2SO_4)	~ 0.3 mL
1	125 mL	Dissolved Metals	10 mL	Nitric (HNO_3)	~0.1 mL
1	40 mL	Dissolved Organic Carbon	25 mL	Sulfuric (H_2SO_4)	~ 0.3 mL
1	40 mL	Dissolved Inorganic Carbon	25 mL	None	-

Note: Pore water samples were separated by bottle size, test, minimum volume required for lab analysis, and preservation type.

Soil characteristics were measured in the field using an Orion Star A329 field probe which was calibrated before each sampling event with pH 4, 7, and 10 calibration solution, ORP calibration standard, and 1413 µS/cm conductivity calibration solution. An area within a foot of each Rhizon sampler was identified as a point of measurement, and a steak knife was used to create an entry point for the probe to be inserted into. Measurements were recorded for soil temperature, pH, dissolved oxygen, oxidation-reduction potential, electrical conductivity, and moisture content at each sampling point along the transect. These data helped to create an abiotic dataset that allows for quantification of nutrient cycling rates and patterns of the restored sites.

2.3.4 Sedimentation

To measure sedimentation, sediment pit trips were installed in the thalweg at the downstream end of each of the 9 sites (Figure 4). A plastic dish rack was used as the base and plastic bottles were used as the containers that captured the sediment; each bottle was approximately 10.25 cm tall, 7.75 cm long, and 7.75 cm wide. The traps held 36 containers set up in a 6x6 grid system. The containers were zip tied to the plastic rack and the apparatus was secured to the stream bed with J-hook rebar. The sediment trap was installed so that it was nested into the stream bed rather than sitting above it for the purpose of maximizing sediment capture.

A random number generator was used to create a sequence that determined which container from the trap would be selected first. The first of two samples was sent to Alloway, a commercial environmental lab in Lima, Ohio. Alloway conducted nutrient analyses on the samples, including total carbon, total nitrogen, total phosphorus, percent total solids, and percent total volatile solids. The sample needed to fill a 4-ounce glass jar, so the number of containers used from each site was determined by how much sediment was deposited in the containers. The sample was extracted from the containers, transferred into a metal bucket, and mixed with a spoon so that a composite sample could be created.

A secondary sample (enough to fill ¼ of a gallon bag) was collected from additional containers. It was brought back to the Ohio University lab for grain size distribution analysis for three size classes: fine (< 425 μm), medium (2 mm $- 425$ μm), and coarse (> 2 mm). The samples were placed into pre-weighed aluminum pans and then transferred into an oven to remove moisture. The oven was set to medium-high heat and typically took around 6 hours to thoroughly dry the sample. Once the samples were dried and cooled, they were each weighed to determine the dry mass. The sample was then passed through two stainless steel sieves (one with 2 mm mesh and the other with 425 μm mesh) to separate the grains into the three size classes. Sediment from each size class was weighed to determine the makeup of the sample. In several cases, the sediment traps were carried away during high flow periods. A stainless steel trowel was used in place of the sediment traps to collect a sample from the middle of the channel since this is where the sediment traps were located. The trowel was triple rinsed between sites to avoid contamination.

2.4 Statistical Analysis of Data

To assess the interactions between these variables, RStudio software was used to complete statistical analysis on the data (R Core Team, 2019). Correlation tests were used to determine pairwise associations between surface water nutrients, pore water nutrients, sediment nutrients, flow rate, and sediment grain size. In every case, a Spearman's Rank Order test was used as a nonparametric alternative due to the size of the dataset. If necessary, a log transformation was used to better grasp the relationship between variables. The results from correlation tests were considered significant if the resulting p-value is less than 0.05, and the strength of the correlation was determined by the correlation coefficient.

Linear regressions were used to further examine relationships between variables. These tests were used to compare flow rate and surface water nutrients (nitrogen and phosphorus) for total, dissolved, and suspended forms by restoration status. Several graphs were produced in R to visualize the normality, variance, and linearity of the data.

A p-value lower than 0.05 identified significant linearity. The R^2 value was reported so that the proportion of variance within the model could be understood.

T-tests were used to determine if there was a significant difference between a numerical variable in two groups. These tests were used to analyze differences between sites by restoration status to determine differences in the sediment grain size, surface water nutrients, pore water nutrients, sediment nutrients, water storage, nutrient export, and surface water nutrient loading. In each assessment, the nonparametric Wilcoxon signed-rank test were used because the data was not normally distributed. A Kruskal-Wallis ANOVA was used to compare between three categorical variables, such as sampling date. The post hoc Dunn's test was used to assess difference among variables tested in the ANOVA. For both tests, results were considered significant if the p-value was less than 0.05. Boxplots assisted these numerical results to provide a visual of the data.

Chapter 3: Results

3.1 Flow

The sampling events in July and November of 2020 occurred during a period of low and base flow respectively. According to the nearby USGS gauge at Wheeling Creek in Elm Grove, West Virginia (~ 18 miles from the furthest restored site), there was no rainfall during the week before sampling in July and about 0.75 inches the week before sampling in November (USGS, n.d.). The March 2021 event occurred after the area received about 1.20 inches of rain the day before sampling began, and this rainfall resulted in higher flow values at all sites compared to the previous two sampling events.

Flow regime differed by stream size class and sampling event. Flow was not different between restored and unrestored streams divided by primary headwaters ($W = 8$, $N = 9$, $p = 0.329$), headwaters ($W = 4$, $N = 9$, $p = 0.461$), and wadeable ($W = 10$, $N = 12$, $p = 1$). The highest flow values were recorded in the largest size class at every sampling event. Figure 5 shows average flow rates for each size class over each sampling event. Dunkard Fork, the largest site, has no flow reading for March 2021 because the water was too high to safely access the site, so that suggests that Dunkard Fork would have also had a relatively high flow value in March. The lowest flow values were found in the primary headwaters size class in July 2020. In November 2020, however, the lowest flow was in Molinari Tributary, a headwaters stream. The lowest flow in March 2021 was found at Beham, which is also a headwaters stream. A linear model was created to see if flow values varied proportionally by drainage area, and it was found that there was no significant linear relationship between the two ($r^2 = 0.097$, $p = 0.11$). Log transformed flow values differed by sampling event. July had significantly lower flow than March (Z = -3.86, p = 0.0001), July had lower flow than November (Z = -2.19, p = 0.03), and March nearly had higher flow than November (Z = 1.96, p = 0.0503).

Figure 5

Flow Rate by Sampling Date and Size Class

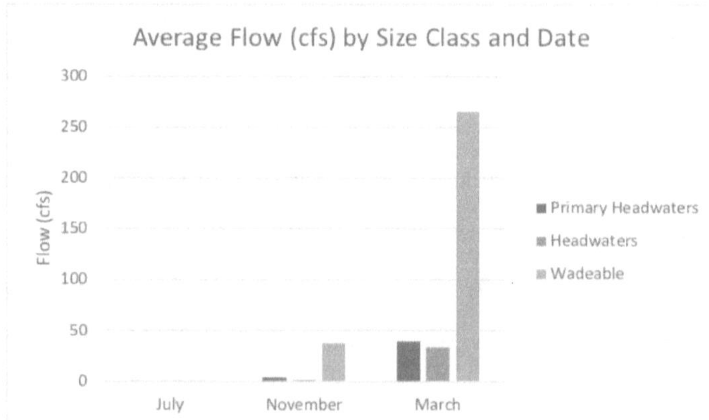

Note: Average flow values (cfs) calculated from in-field salt tracing by site size and date. The highest flow values in each month corresponded with the largest stream size class. In July, flow ranged from 0.00064-0.154 cfs where the lowest average flow was found at the primary headwaters sites.

Precipitation data, collected from a nearby National Weather Service gauge station (NWS, n.d.), is shown from 2012-2021 in Figure 6 for reference. Lebanik, the largest restored wadeable stream by drainage area, exhibited less variation in water depth during the two years of available post-restoration data (Figure 7). The hydrograph for

Molinari, also a wadeable stream, shows that the average water depth is similar from pre

to post-restoration, but change in water depth is much less variable after restoration

(Figure 8). Beham, a headwaters stream, showed a wider range of water depth after

restoration than it did before (Figure 9). However, Molinari Tributary, the other

headwaters stream, had a low average water depth in 2018, which was the wettest year

based on total annual rainfall. The 2017 and 2019 data show little variability in water

depth (Figure 10). McCully, a primary headwaters stream, had overall smaller ranges in

precipitation post-restoration, although the change is less apparent (Figure 11). No pre-

restoration water depth data was available at Unit 4D, the other primary headwaters site

(Figure 12). A more in-depth analysis of the hydrology at these study sites can be found

in Pazol (2021).

Figure 6

Precipitation at Waynesburg, PA

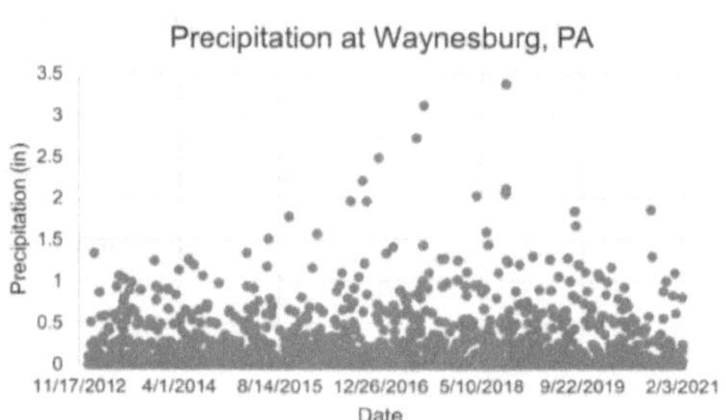

Note: Daily precipitation data at the Waynesburg, Pennsylvania 1E Gauge Station, collected by the National Weather Service. 2018 had the highest amount of total annual rainfall.

Figure 7

Lebanik Water Depth Over Time

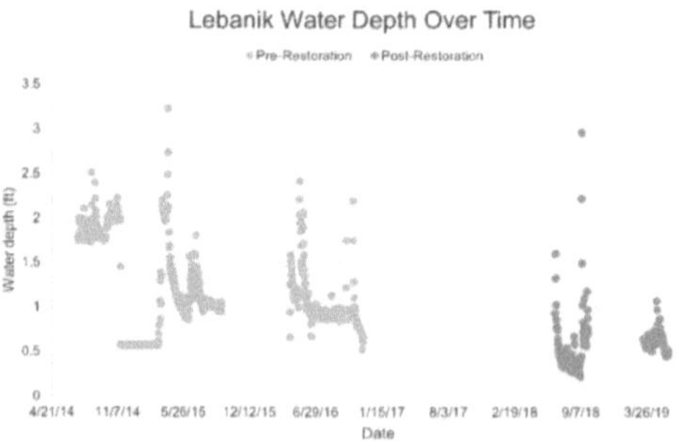

Note: Water depth data by restoration status at Lebanik, a wadeable stream. The range in water depth generally became smaller after restoration occurred.

Figure 8

Molinari Water Depth Over Time

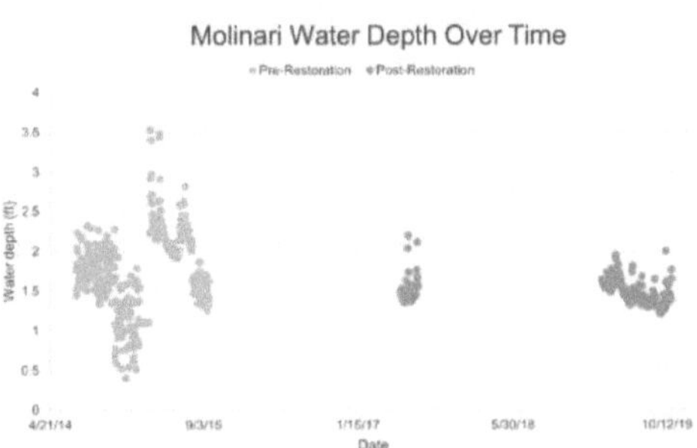

Note: Water depth by restoration status at Molinari, a wadeable stream. Variation in water depth had a smaller range after restoration.

Figure 9

Beham Water Depth Over Time

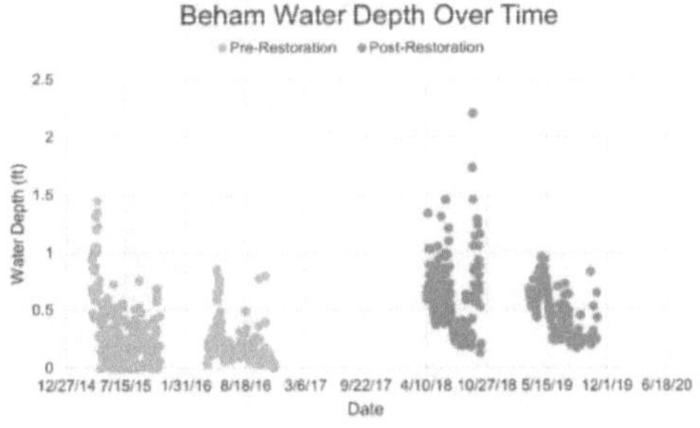

Note: Water depth by restoration status at Beham, a headwaters stream. The post-restoration range in water depth was more variable than it was pre-restoration, contrary to trends at the other sites.

Figure 10

Molinari Tributary Water Depth Over Time

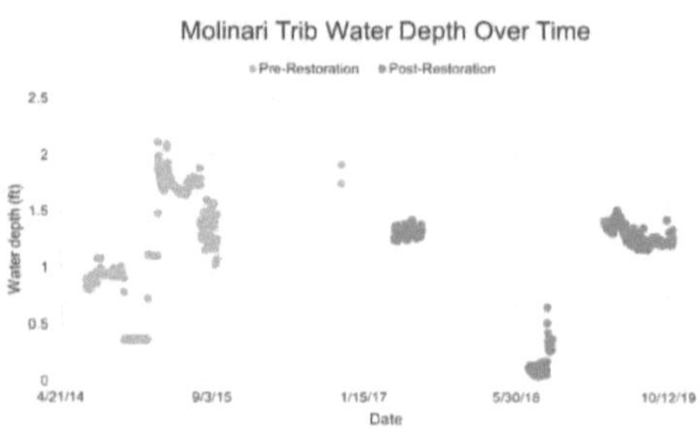

Note: Water depth over time at Molinari Tributary, a headwaters stream. Water depth was low in 2018, which was the year with the highest amount of rainfall. Water depth ranges post-restoration were less variable than pre-restoration.

Figure 11

McCully Water Depth Over Time

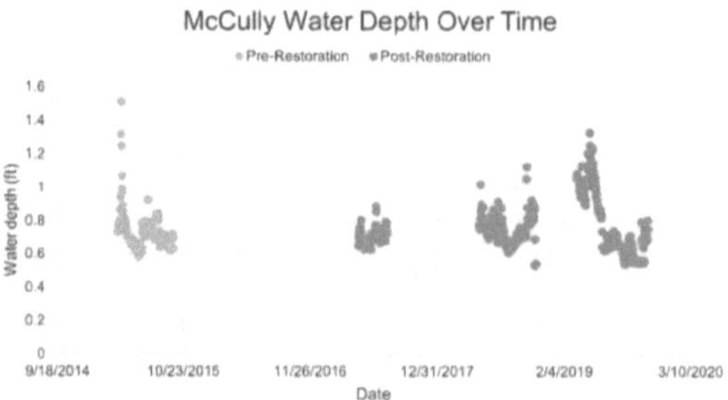

Note: Water depth by restoration status at McCully, a primary headwaters stream. There was slightly less variation in water depth post-restoration.

Figure 12

Unit 4D Water Depth Over Time

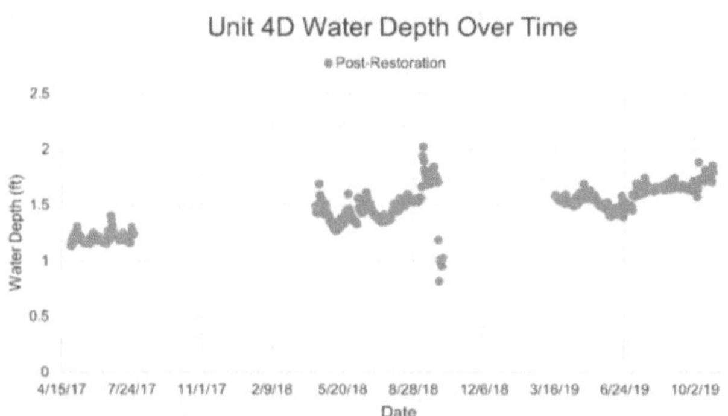

Note: Water depth over time at Unit 4D, a primary headwaters stream. Pre-restoration data was unavailable at this site.

Water storage was calculated by subtracting the channel flow reading from the salt tracing flow measurement. This calculation was based on the assumption that the salt tracing flow measurement would include short term transient water storage from the vadose zone. Over the three sampling events, post-restoration sites stored water (i.e., tracer flow > channel flow) 71.43% of the time and did not store water (i.e., tracer flow < channel flow) 28.57% of the time. Pre-restoration sites stored water 66.67% of the time and did not store water 33.33% of the time. Water storage was not significantly different between restored and unrestored sites (W = 50, N = 20, p = 0.55). There was no significant difference between water storage based on the sampling event, even though flow was highest at all sites during the March 2021 event ($\chi^2(2) = 4.75$, p = 0.09). There was also no difference between water storage based on stream size class ($\chi^2(2) = 2.84$, p = 0.24). These results suggest that there was not a significant increase in short term transient water storage at the restored sites, but long-term storage should be quantified in future data collection and analysis.

3.2 Sediment

Sediment grain size distribution was analyzed to determine if grain size differed by restoration status (Figures 13, 14). There was no difference in coarse (> 2 mm) or medium (425 μm – 2 mm) sediment composition based on restoration status (W = 15, N = 17, p = 0.25; W = 30, p = 0.7 respectively). Conversely, there was a trend toward more fine sediment in the post-restoration sites, but the difference was insignificant (W = 42, p = 0.08). Flow and sediment dynamics were also considered so that mechanisms of sediment transfer could be identified. The relationship between flow and sediment grain size was analyzed for each size class; as flow increased, the proportion of fine sediment in the sample increased as well (r = 0.59, p = 0.02) (Figure 15). There was no correlation between flow rate and medium sediment (r = -0.15, p = 0.57) or coarse sediment (r = -0.26, p = 0.31).

Figure 13

Sediment Grain Size Distribution at Unrestored Streams

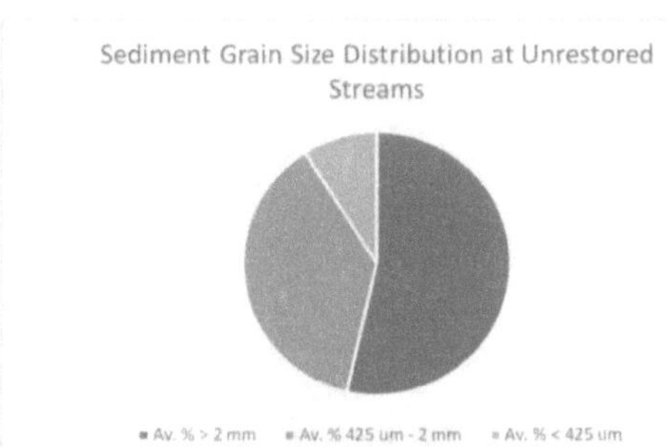

Note: Average percent of each grain size class at unrestored streams. The coarse sediment

size class (> 2 mm) made up over half of the sample on average.

Figure 14

Sediment Grain Size Distribution at Restored Streams

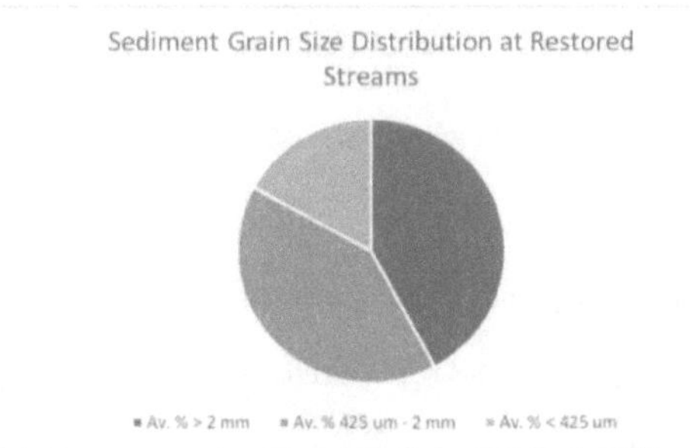

Note: Average percent of each grain size at restored streams. The coarse (> 2 mm) and

medium (425 μm – 2 mm) size classes dominated the sample on average. Fine sediment

made up a larger proportion of the sample at restored sites than unrestored.

Figure 15

Flow and Percent Fine Grain Sediment

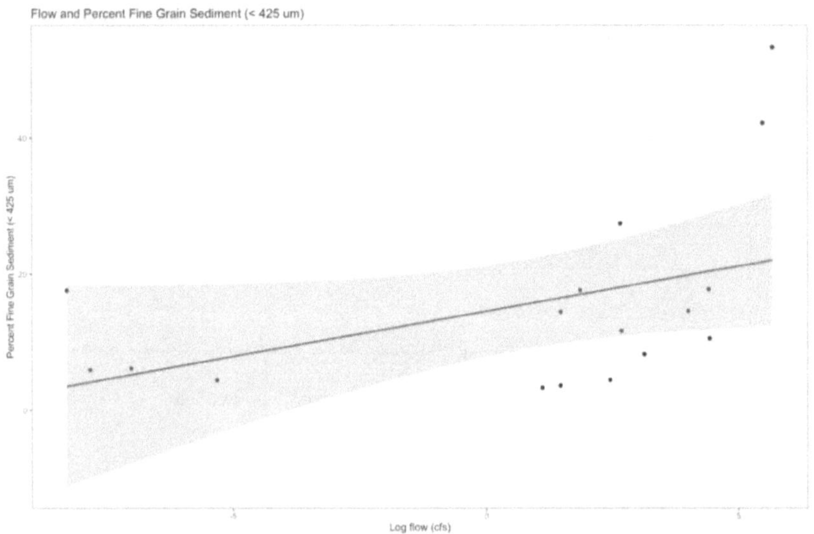

Note: Percent fine grain sediment (< 425 μm) was plotted against log-transformed flow. The proportion of fine grain sediment within the sample increased with flow rate (r = 0.59, p = 0.02). The confidence interval is shown by the shaded gray area.

Total suspended solids (TSS) concentrations were variable by site and sampling event (Figure 16). Overall, TSS concentrations were highest in July, but loads were highest in March due to the influence of flow rate on load calculation. In more than half of the samples, TSS load was less than 5 mg/L. The remainder of the samples had a TSS load from 100-6000 mg/L. The variation in load cannot be clearly explained by restoration status, but flow did have an impact on downstream TSS load. Flow, scaled by a log-transformation, was correlated with an increase in downstream TSS load ($r^2 = 0.3$, $N = 21$, $p = 0.01$) (Figure 17). However, the relationship appears to be driven by only a few data points. There was no significant difference between downstream TSS load by restoration status ($W = 66$, $p = 0.12$) or by sampling event ($\chi^2(2) = 2.51$, $p = 0.17$) even though downstream TSS loads were higher in March (Figure 18).

Figure 16

TSS Concentration (mg/L) by Site and Month

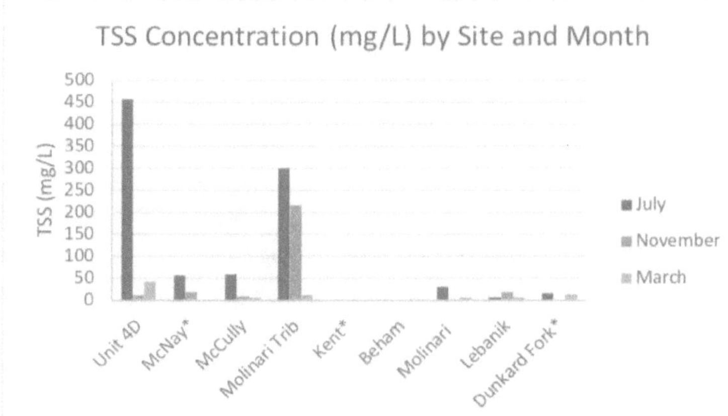

Note: TSS concentrations (mg/L) were variable across site and sampling event. The sites increase in drainage area from left to right. Where there is no visible TSS concentration, TSS was found to be below the detection limit. Unrestored sites are labeled with an asterisk after their name.

Figure 17

TSS Load (mg/L) by Site and Month

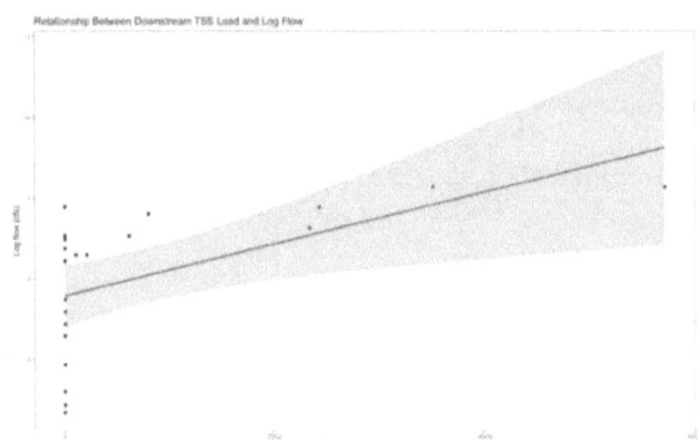

Note: Downstream TSS load, a measurement of export, was driven by flow ($r^2 = 0.3$, p = 0.01).

Figure 18

Relationship Between Downstream TSS Load and Log Flow

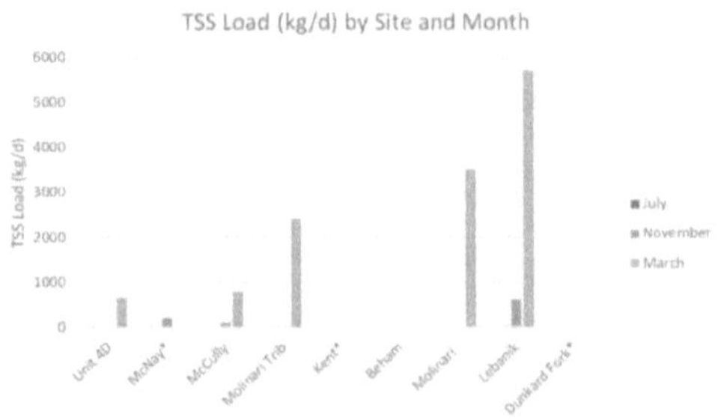

Note: Downstream TSS load was low in July and November 2020 compared to March 2021. In March, TSS loads were lowest at Beham and the three unrestored sites, McNay, Kent, and Dunkard Fork.

Sediment deposition in kilograms per day was calculated using the change in TSS loading from upstream to downstream, where negative values represent resuspension into the water column (Figure 19). A linear regression was performed to determine if log-transformed flow drives sediment deposition, and it was found that a low flow rate does not suggest sediment deposition ($r^2 = 0.08$, N = 12, p = 0.19). Sediment deposition per linear foot was low across all sampling events ranging from -0.2—2.0 for all restored sites. TSS export per square mile was analyzed by restoration status, and there was no difference between restored and unrestored sites (W = 63, N = 21, p = 0.18) (Figure 20).

Figure 19

Sediment Deposition by Sampling Event

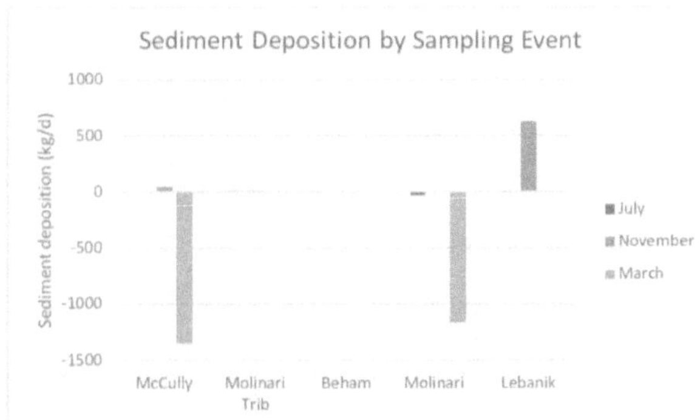

Note: Sediment deposition was depicted by the change in TSS load from upstream to downstream where negative values mean TSS was being retained and positive values mean it was being exported. Data are available for sites with an upstream and downstream site. Upstream data were not collected at Beham in July or November.

Figure 20

TSS Load Export per Square Mile

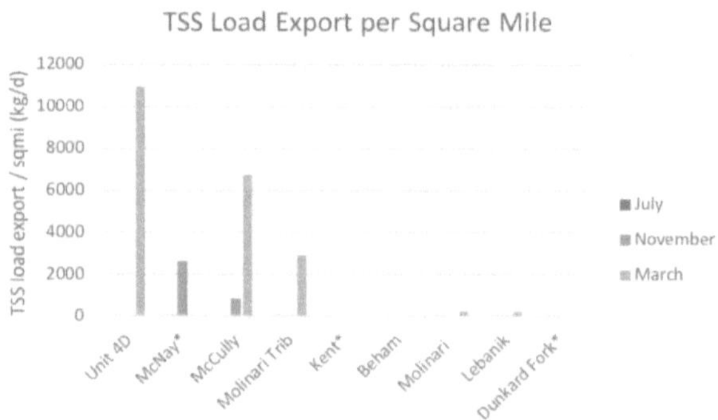

Note: TSS load export per square mile was lowest in July 2020 during a period of low flow. Export was highest in March 2021. The restored sites showed low amounts of TSS export per square mile.

3.3 Nutrients

3.3.1 Sediment Nutrients

Sediment nutrient concentrations derived from samples collected within the channel (Figures 21, 22) were compared by restoration status to identify potential differences. There was significantly more total nitrogen in the sediment in the restored sites compared to the unrestored ($W = 99$, $N = 22$, $p < 0.001$) (Figure 23). Similarly, there was significantly more total phosphorus in the sediment in restored streams compared to unrestored ($W = 91$, $p = 0.008$) (Figure 24). Sediment at restored sites was richer in nutrients than it was at unrestored sites, offering evidence of the impact of restoration.

The relationship between flow and sediment nutrient data was also compared to identify any interaction. There was no significant correlation between flow and sediment nitrogen concentration (r = -0.24, N = 17, p = 0.36) or sediment phosphorus concentration (r = 0.02, p = 0.94). This result suggests that flow and sediment nutrient interactions were not significant at these study sites.

Figure 21

Sediment Nitrogen Concentrations

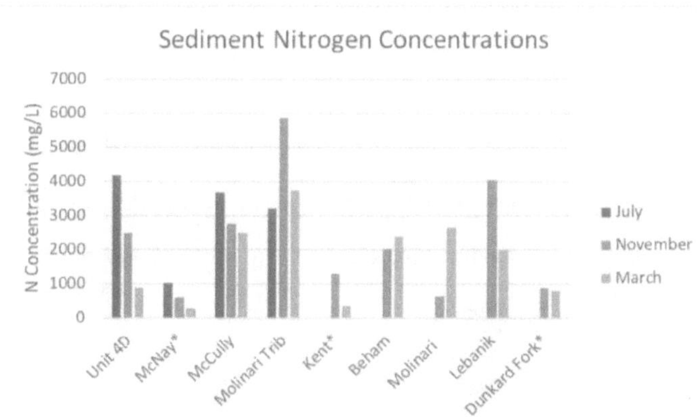

Note: Sediment nitrogen concentrations were unavailable at several sites in July due to missing sediment traps. Molinari Tributary, a headwaters stream saw the highest concentrations of nitrogen in both November and March. The unrestored sites had lower nitrogen concentrations.

Figure 22

Sediment Phosphorus Concentrations

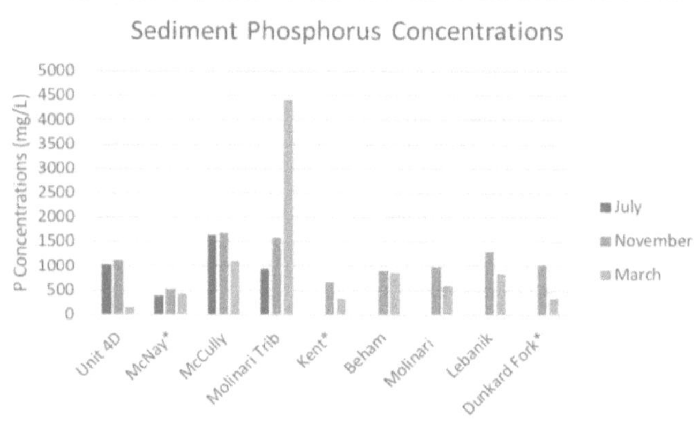

Note: Sediment phosphorus concentrations were unavailable at several sites in July due to missing sediment traps. Molinari Tributary saw the second highest phosphorus concentration in November and the highest in March. The unrestored sites had lower phosphorus concentrations.

Figure 23

Sediment Total N Concentration by Restoration Status

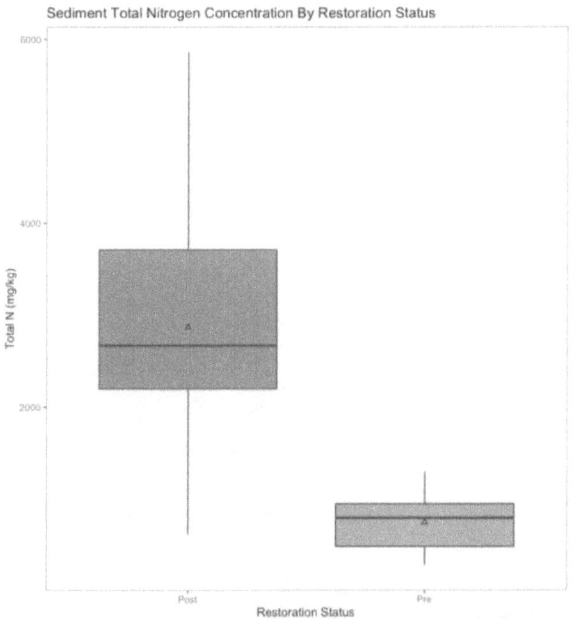

Note: Sediment total nitrogen concentrations were significantly higher in post-restoration sites (N = 22, p < 0.001).

Figure 24

Sediment Total P Concentration by Restoration Status

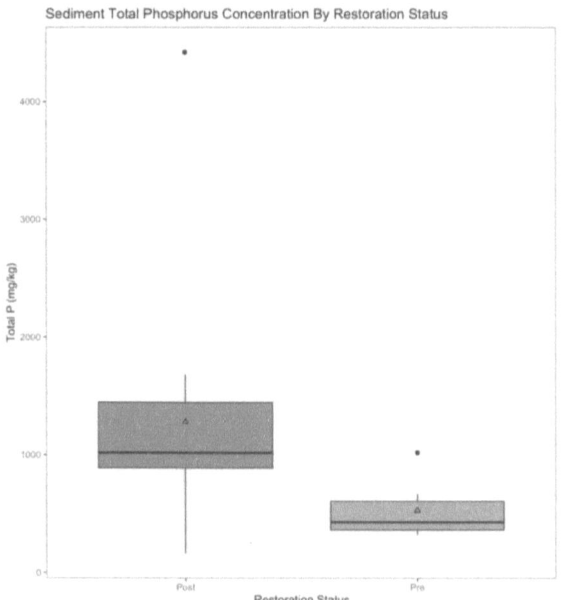

Note: Sediment total phosphorus concentrations were significantly higher in post-restoration sites (N = 22, p = 0.008).

3.3.2 Surface Water Nutrients

Surface water nutrient concentrations were variable by site and date. Table 5 shows the concentrations of nitrogen and phosphorus, partitioned into total, dissolved, and suspended forms where suspended is the difference between total and dissolved. After performing a log transformation on flow, no significant correlation was found between flow and surface water downstream total nitrogen concentration ($r = -0.039$, $N = 28$, $p = 0.87$) (Figure 25). Similarly, no correlation was identified between the log transformed flow and surface water downstream total phosphorus concentration ($r = -0.32$, $p = 0.16$) (Figure 26). The same analysis was performed on the relationship between log transformed flow and dissolved nutrient concentrations, and there was no significant correlation for nitrogen or phosphorus ($r = 0.24$, $p = 0.3$; $r = -0.031$, $p = 0.9$ respectively). There was no significant correlation between log transformed flow and suspended nitrogen or phosphorus ($r = -0.28$, $p = 0.22$; $r = -0.29$, $p = 0.2$ respectively). However, a trend between flow and nutrient concentrations is apparent when the data were graphed. For all nutrient forms, high flow seems to relate to low nutrient concentrations.

Table 5

Surface Water Nutrient Concentrations

Site	Total N	Diss. N	Susp. N	Total P	Diss. P	Susp. P	Date	Location
Lebanik	<0.25 (U)	<0.0972 (U)	0	0.035	0.033	0.002	July	Upstream
Lebanik	<0.25 (U)	0.197	0	0.024	0.019	0.005	July	Downstream
Beham	0.34	0	0.34	0.061	0.042	0.019	July	Downstream
Molinari Trib.	0.51	0.377	0.133	0.143	0.05	0.093	July	Upstream
Molinari Trib.	1.01	0.282	0.728	0.128	0.027	0.101	July	Downstream
Molinari	0.32	0.158	0.162	0.054	0.018	0.036	July	Upstream
Molinari	0.52	0.399	0.121	0.085	0.033	0.052	July	Downstream
McCully	1.25	0.034	1.216	0.131	0.028	0.103	July	Upstream
McCully	0.47	0.32	0.15	0.075	0.019	0.056	July	Downstream
Unit 4D	1.52	1.75	0	0.1	0.06	0.04	July	Downstream
McNay	0.49	0.436	0.054	0.081	0.065	0.016	July	Downstream
Kent	<0.25 (U)	<0.0972 (U)	0	0.027	0.026	0.001	July	Downstream
Dunkard Fork	<0.25 (U)	0.25	0	0.077	0.059	0.018	July	Downstream
Lebanik	<0.25 (U)	0.153	0	0.027	0.021	0.006	November	Upstream

Table 5: continued

Lebanik	<0.25 (U)	0.182	0	0.041	0.022	0.019	November	Downstream
Beham	<0.25 (U)	0.235	0	0.031	0.017	0.014	November	Downstream
Molinari Trib.	0.26	0.238	0.022	0.122	0.051	0.071	November	Upstream
Molinari Trib.	2.06	0.251	1.809	0.202	0.018	0.184	November	Downstream
Molinari	<0.25 (U)	0.26	0	0.033	0.029	0.004	November	Upstream
Molinari	<0.25 (U)	0.171	0	0.032	0.022	0.01	November	Downstream
McCully	<0.25 (U)	0.226	0	0.01	0	0.01	November	Upstream
McCully	<0.25 (U)	0.17	0	0.025	0.019	0.006	November	Downstream
Unit 4D	0.79	0.742	0.048	0.09	0.04	0.05	November	Downstream
McNay	0.42	0.267	0.153	0.059	0.049	0.01	November	Downstream
Kent	<0.25 (U)	0.151	0	0.022	0.02	0.002	November	Downstream
Dunkard Fork	<0.25 (U)	0.233	0	0.044	0.03	0.014	November	Downstream

Table 5: continued

Lebanik	0.59	0.668	0	0.039	0.02	0.019	March	Upstream
Lebanik	0.58	0.598	0	0.035	0.019	0.016	March	Downstream
Beham	0.4	0.441	0	0.027	0.019	0.008	March	Upstream
Beham	0.43	0.39	0.04	0.026	0.018	0.008	March	Downstream
Molinari Trib.	0.69	0.605	0.085	0.049	0.035	0.014	March	Upstream
Molinari Trib.	0.55	0.531	0.019	0.054	0.04	0.014	March	Downstream
Molinari	0.64	0.663	0	0.048	0.02	0.028	March	Upstream
Molinari	0.65	0.65	0	0.044	0.21	0	March	Downstream
McCully	0.99	0.951	0.039	0.07	0.034	0.036	March	Upstream
McCully	0.83	0.77	0.06	0.039	0.02	0.019	March	Downstream
Unit 4D	2.37	2.265	0.105	0.105	0.047	0.058	March	Downstream
McNay	0.31	0.318	0	0.031	0.021	0.01	March	Downstream
Kent	0.44	0.451	0	0.03	0.018	0.012	March	Downstream
Dunkard Fork	0.68	0.638	0.042	0.056	0.021	0.035	March	Downstream

Note: Surface water nutrient concentrations (mg/L) are shown by site, month of sampling, and downstream or upstream location when available. Concentrations below the detection limit are notated as (U) following the detection limit.

Figure 25

Flow and N Concentrations by Form

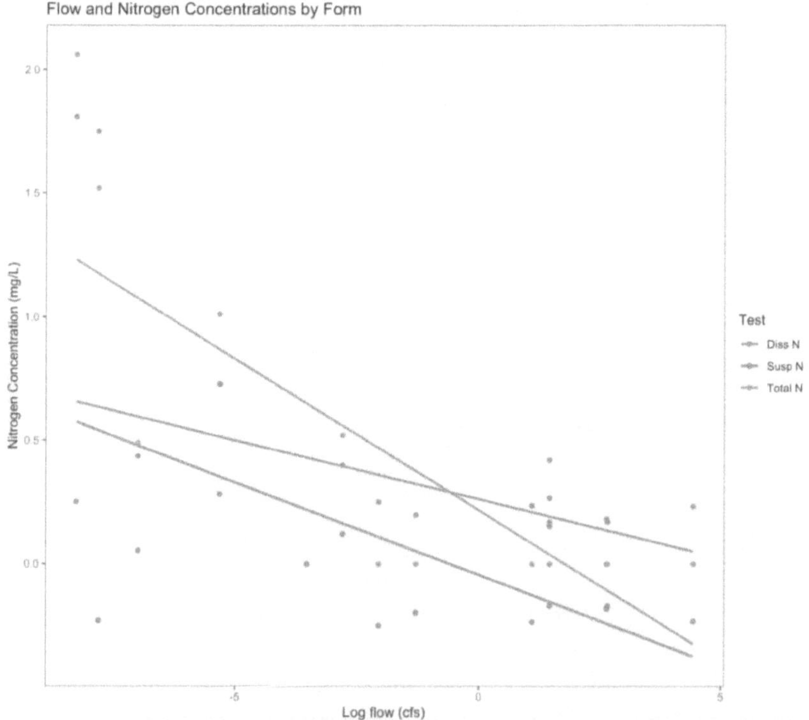

Note: Log-transformed flow and downstream nitrogen concentrations are shown by nitrogen form. Higher flow appeared to correspond with lower nitrogen concentrations for all forms.

Figure 26

Flow and P Concentrations by Form

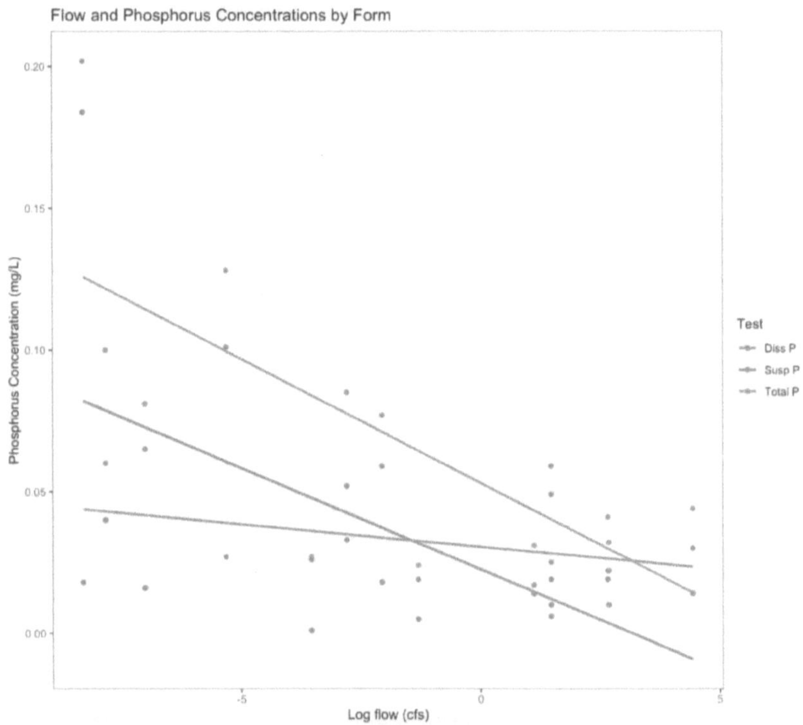

Note: Log transformed flow and downstream phosphorus concentrations are shown by phosphorus form. Higher flow values appeared to correspond with lower phosphorus concentrations for all forms.

Surface water nitrogen to phosphorus ratios were calculated at all sites for total, dissolved, and suspended forms. Figure 27 shows ratios for restored and unrestored sites by form. Total, dissolved, and suspended N:P ratios were analyzed by restoration status to see if restored sites exhibited significantly different ratios than unrestored sites. There was no significant difference between total or suspended N:P by restoration status ($W = 63$, $N = 28$, $p = 0.17$; $W = 47$, $p = 0.9$). There was a significant difference between dissolved N:P ratio by restoration status, where the ratio was higher in restored sites ($W = 75$, $p = 0.018$) (Figure 28). There was no correlation between total N:P ratio and log-transformed flow, dissolved N:P ratio and log-transformed flow, or suspended N:P ratio and log-transformed flow ($r = 0.13$, $p = 0.52$; $r = 0.22$, $p = 0.26$; $r = -0.33$, $p = 0.09$ respectively). These relationships are depicted in Figure 29. Here, it appears that flow may have a positive effect on dissolved N:P and a negative effect on suspended N:P. This suggests that flow has a differing relationship with N:P ratios by form.

Figure 27

N:P Ratio by Form at Restored and Unrestored Sites

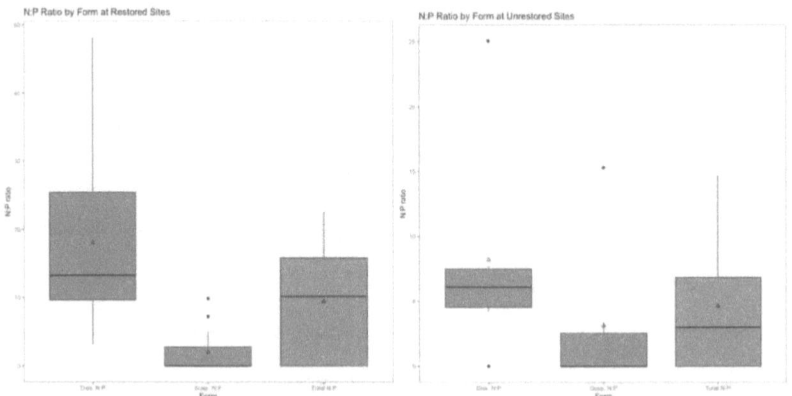

Note: Dissolved, suspended, and total N:P ratios are shown for restored sites on the left and unrestored on the right. Dissolved N:P made up a larger proportion of total N:P at the restored sites. At the unrestored sites, the makeup was similar.

Figure 28

Dissolved N:P by Restoration Status

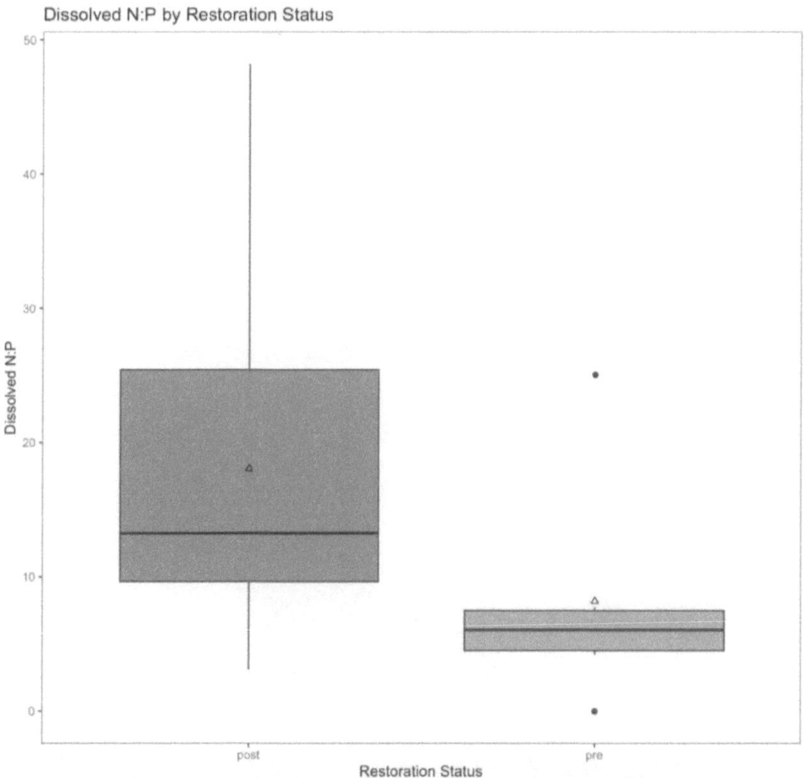

Note: Dissolved N:P ratio was higher at post-restoration sites (W = 75, p = 0.018).

Figure 29

Flow and N:P Ratio by Form

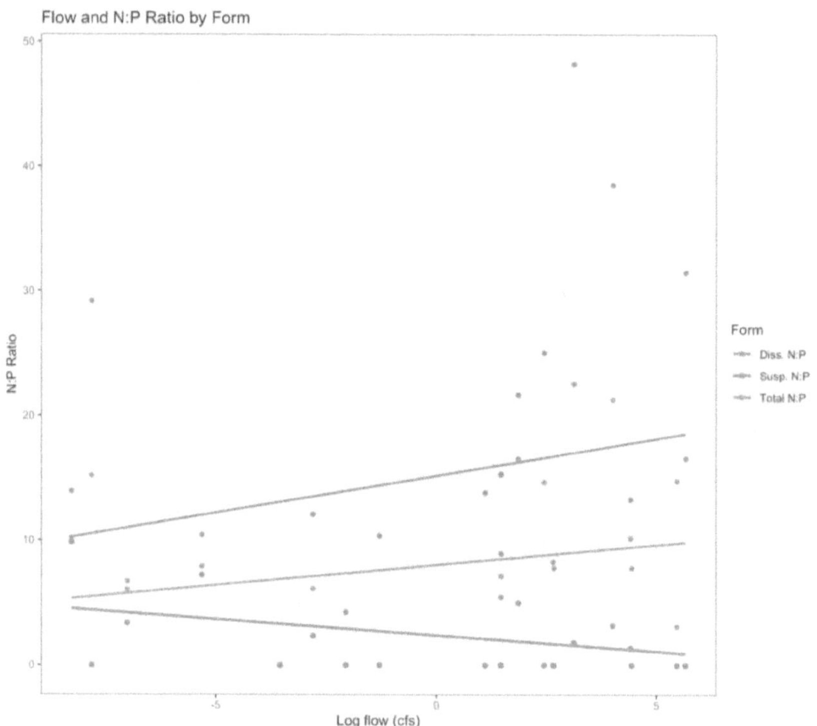

Note: Log flow and N:P ratios were plotted by nutrient form. Dissolved N:P ratio appeared to increase as flow increased. Suspended N:P ratio appeared to decrease as flow increased.

Downstream nutrient loading was analyzed to see if restored sites were acting as nutrient sinks. No significant difference was found between downstream surface water nutrient loading and restoration status for total nitrogen (W = 59, N = 21, p = 0.3) or total phosphorus (W = 52, p = 0.62). The same analysis was tested on dissolved nitrogen and dissolved phosphorus loads, and neither yielded significant differences by restoration status (W = 41, p = 0.8001; W = 52, p = 0.62). Similarly, there was no significant difference between suspended nitrogen load (W = 43, p = 0.66) or suspended phosphorus load by restoration status (W = 52, p = 0.62) (Figures 30, 31). Downstream nutrient loading appeared to be generally unimpacted by restoration, but it increased with each sampling event, suggesting that there may be a relationship between loading and flow. Both total nitrogen and total phosphorus downstream loading appeared to increase with the increase of flow rate (r^2 = 0.28, p = 0.01; r^2 = 0.36, p = 0.004). Total nitrogen and phosphorus loading can be visualized from upstream to downstream at restored sites in Figures 32, 33, 34, 35, 36, and 37.

Figure 30

Downstream N Load by Date and Form

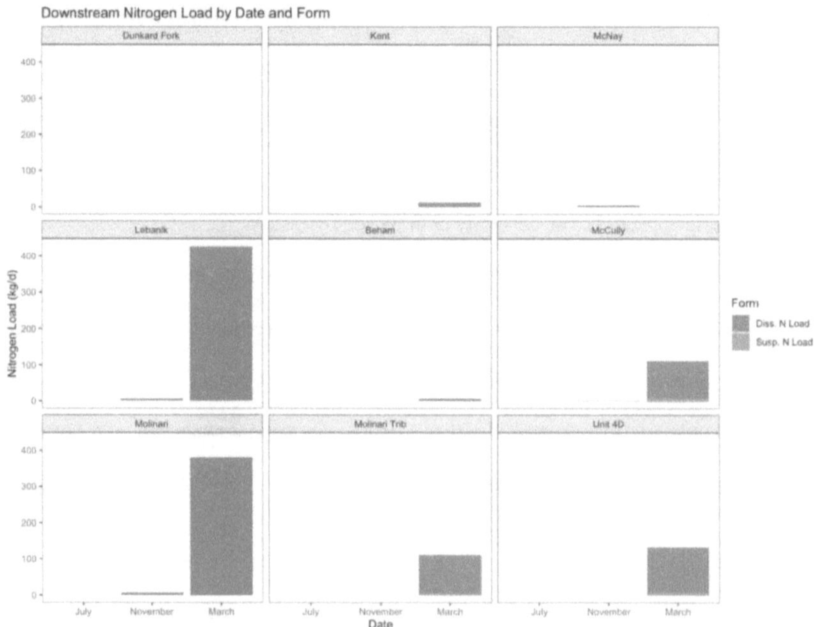

Note: Downstream nitrogen load was lowest at the unrestored sites, shown on the top

row. The wadeable streams, shown in the first column, generally had the highest

downstream loads. In most cases, the highest loading occurred in March.

Figure 31

Downstream P Load by Date and Form

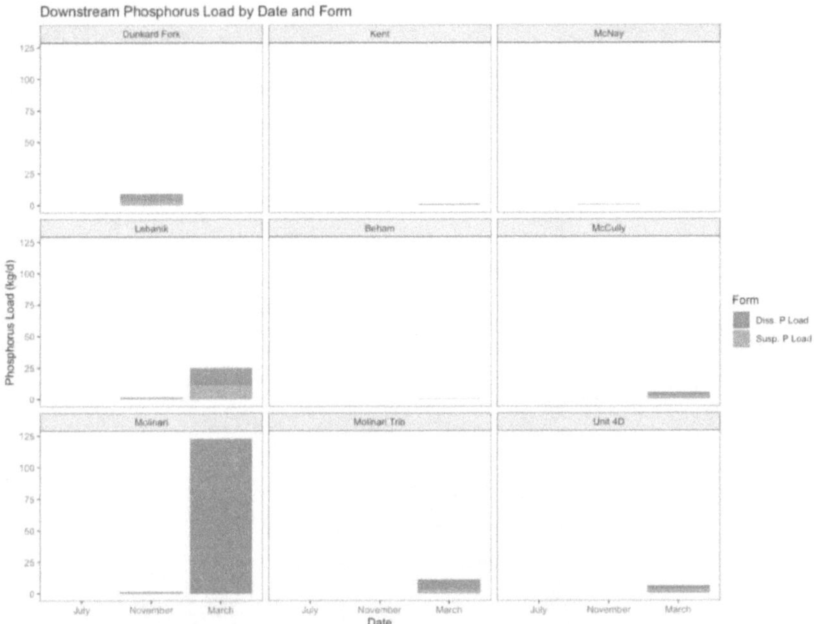

Note: Downstream phosphorus load was overall lowest at the unrestored sites, shown on

the top row. The wadeable streams, shown in the first column, generally had the highest

downstream loads. In most cases, the highest loading occurred in March.

Figure 32

July 2020 N Load by Location and Form

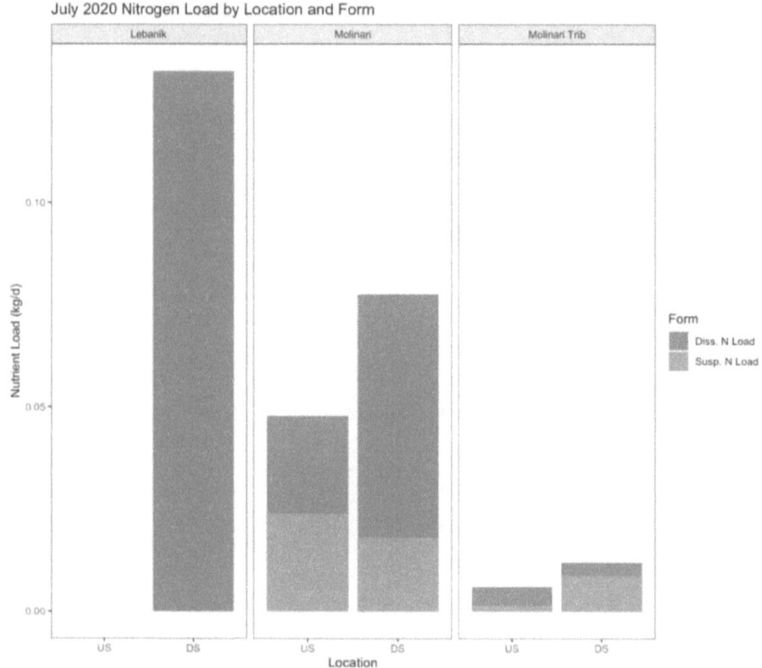

Note: Nitrogen load is shown from upstream to downstream at restored sites in July where flow and nitrogen concentrations were available. Dissolved load typically outweighed suspended load.

Figure 33

November 2020 N Load by Location and Form

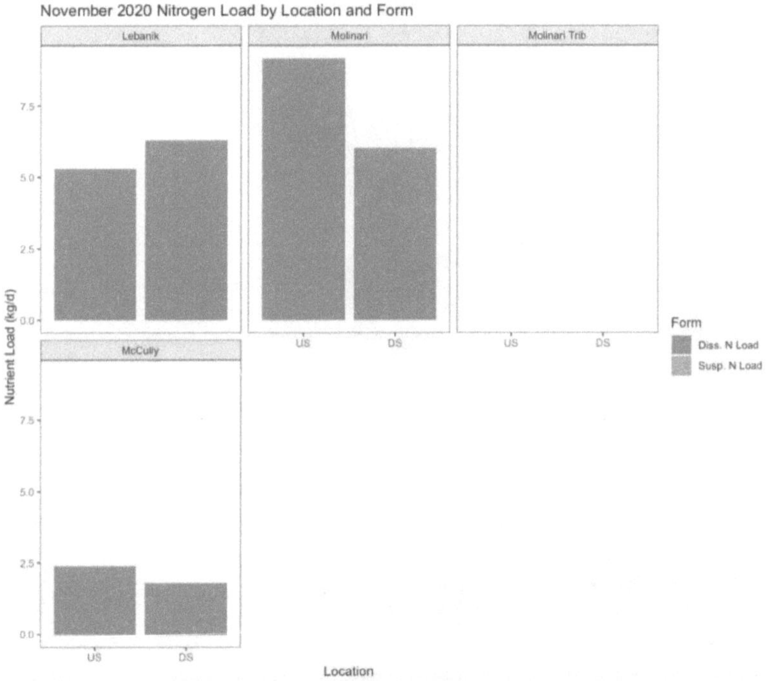

Note: Nitrogen load is shown from upstream to downstream at restored sites in November where flow and total nitrogen concentrations were available. Suspended nitrogen load was nonexistent except for a small amount at Molinari Tributary (< 0.01 kg/d).

Figure 34

March 2021 N Load by Location and Form

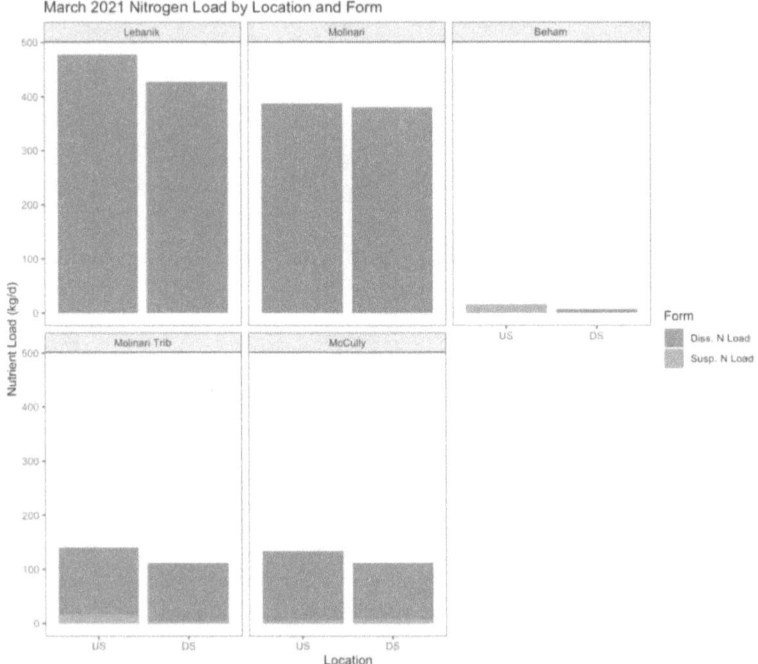

Note: Nitrogen load is shown from upstream to downstream at restored sites in March where flow and total nitrogen concentrations were available. Dissolved nitrogen load outweighs suspended load except upstream at Beham.

Figure 35

July 2020 P Load by Location and Form

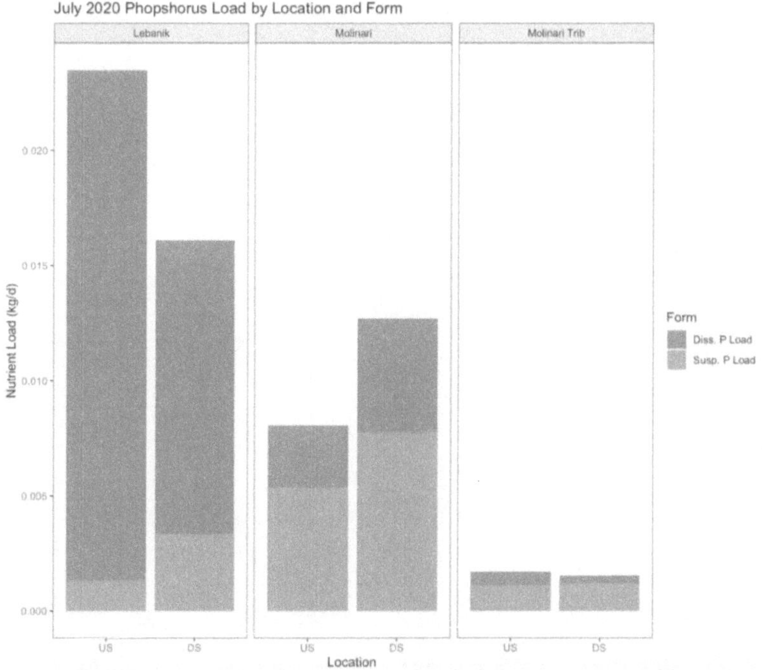

Note: Phosphorus load is shown from upstream to downstream at restored sites in July where flow and total phosphorus concentrations were available. Lebanik loads were dominated by dissolved phosphorus, Molinari loads were dominated by suspended phosphorus, and Molinari Tributary loads were relatively similar in makeup.

Figure 36

November 2020 P Load by Location and Form

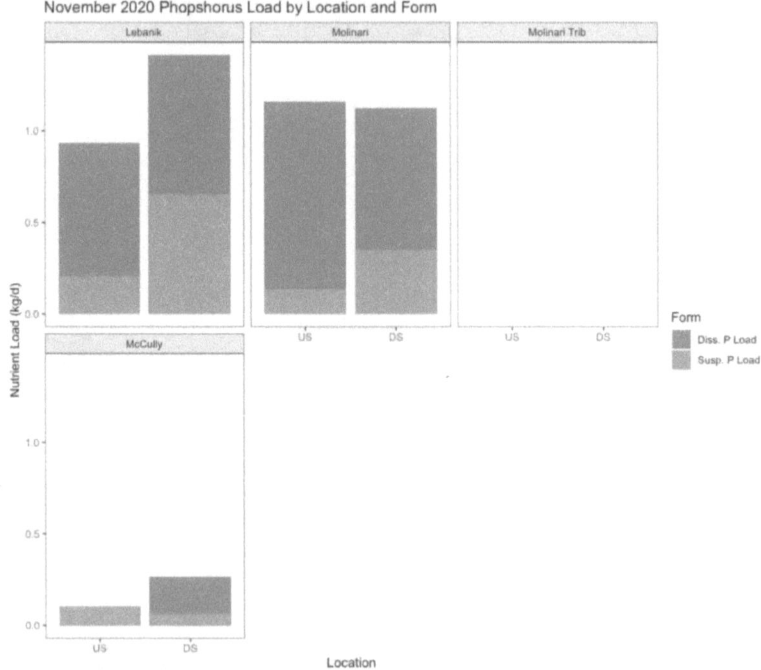

Note: Phosphorus load is shown from upstream to downstream at restored sites in November where flow and total phosphorus concentrations were available. Phosphorus loads at Molinari Tributary were < 0.001 kg/d. Dissolved phosphorus loads were generally dominant.

Figure 37

March 2021 P Load by Location and Form

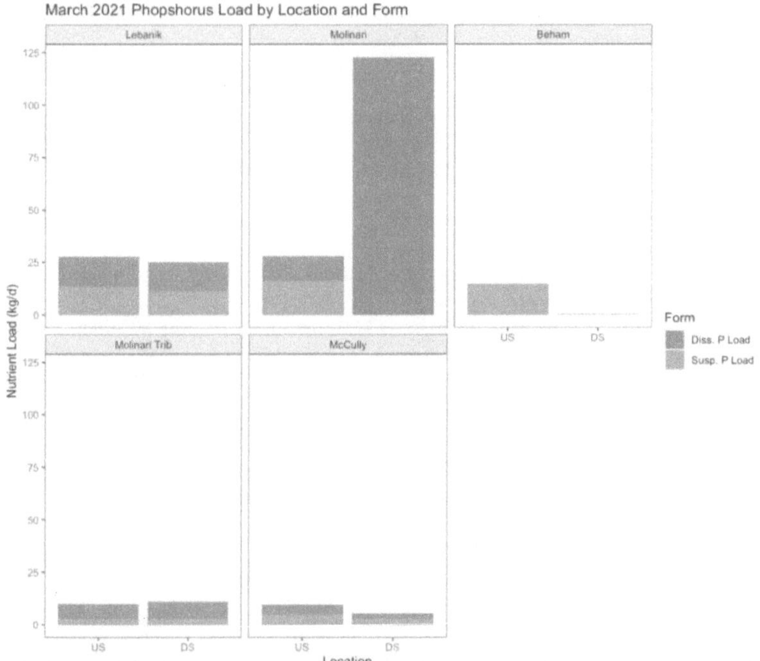

Note: Phosphorus load is shown from upstream to downstream at restored sites in March where flow and total phosphorus concentrations were available. Phosphorus load makeup is similar except for a lack of suspended load at Molinari downstream. Beham upstream is dominated by suspended load and downstream is mostly comprised of dissolved load.

Downstream nutrient loads were assessed as a measure of nutrient export. Nutrient export per square mile is helpful for understanding the presence of upstream nutrient sources as well as nutrient retention or output. The relationship between surface water nutrient export per square mile and restoration status was also investigated to determine if nutrient export was lower in restored sites. There was no significant difference for total nitrogen or total phosphorus export between restored and unrestored sites (W = 54, N = 21, p = 0.52; W = 51, p = 0.68 respectively). Unrestored sites generally had lower export per square mile, but in November, McNay had high amounts of both nitrogen and phosphorus export per square mile. The largest streams tended to have the smallest amount of export per square mile. Nutrient export per square mile is shown in Figures 38, 39, and 40 for July, November, and March respectively.

Figure 38

July Nutrient Export per Square Mile

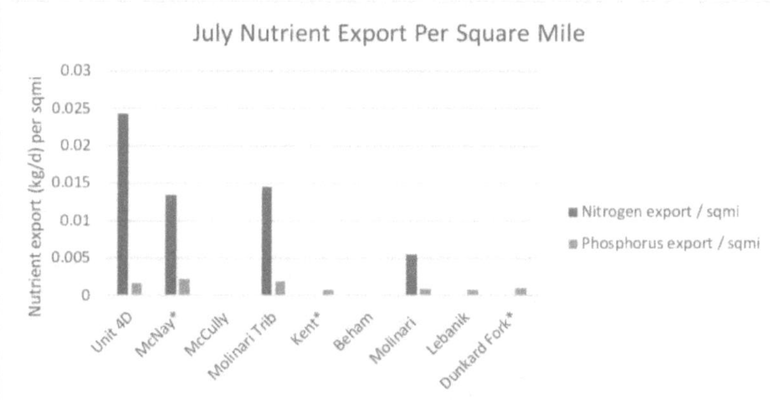

Note: July total nitrogen and total phosphorus export per square mile is shown for the sites where flow data was available. Nitrogen export was undetected where total nitrogen concentrations were also undetected. Phosphorus export was relatively low at all sites. The unrestored sites, McNay, Kent, and Dunkard Fork showed similar export values compared to the restored sites.

Figure 39

November Nutrient Export per Square Mile

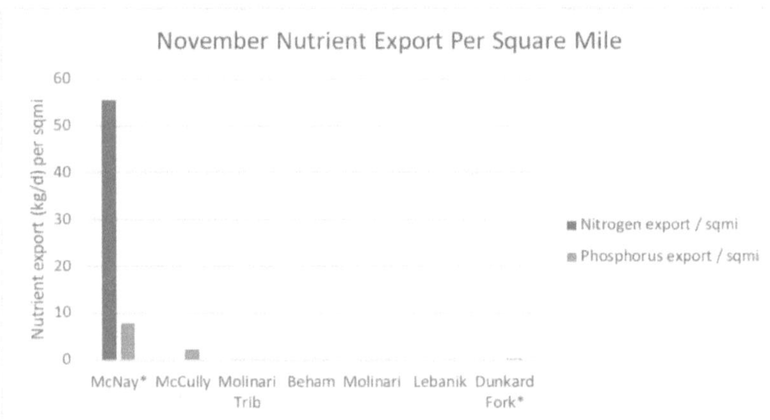

Note: November total nitrogen and total phosphorus export per square mile is shown for the sites where flow data was available. Both nitrogen and phosphorus export were highest at McNay, a primary headwaters unrestored site. Nitrogen concentrations were mostly undetected in November, and phosphorus concentrations were also very low.

Figure 40

March Nutrient Export per Square Mile

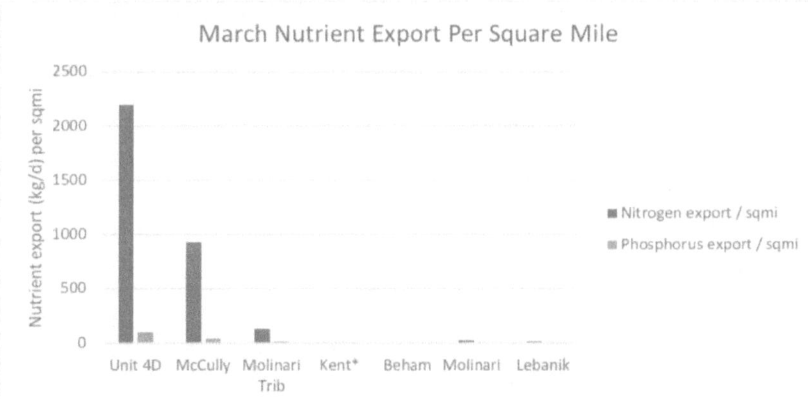

Note: March total nitrogen and phosphorus export per square mile is shown for all sites besides Dunkard Fork and McNay, where flow data was unavailable. Nutrient export was highest at Unit 4D, the smallest of any site by drainage area. McCully, also a primary headwaters stream had the second highest amount of nutrient export.

Change in surface water nutrient load per linear foot was calculated where flow data were available. Negative values show that nutrients are being retained while positive values show that nutrients are being exported. Figures 41, 42, and 43 show change in nutrient load per linear foot in July, November, and March respectively. Retention and export were variable based on sampling date, restoration status, and stream size. Nutrient retention appeared to be greatest for both nitrogen and phosphorus in March 2021. Denitrification rate was calculated as the change in nitrogen load from upstream to downstream. During low flow, more nitrogen was observed downstream compared to upstream, so denitrification was not occurring. In March, denitrification occurred at the two sites that exhibited median flow values. This finding suggests that denitrification is not likely during low flow or high flow and that baseflow conditions might lead to a higher rate of denitrification.

Figure 41

July Change in Nutrient Load per Linear Foot

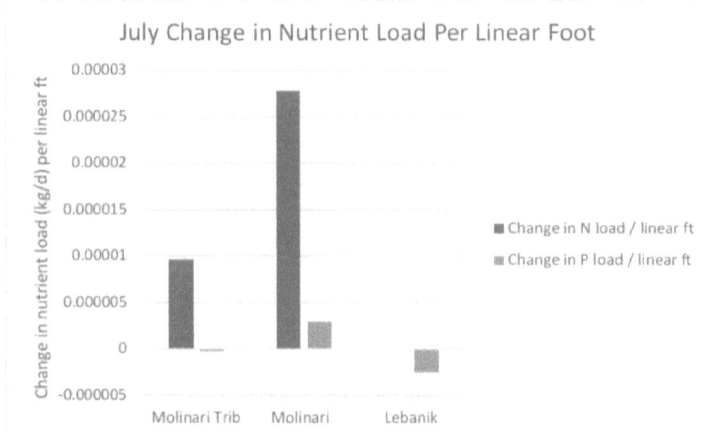

Note: Total nitrogen was exported at Molinari Tributary and Molinari, visualized by the positive bars in July 2020. Phosphorus is being retained at Molinari Tributary and Lebanik, visualized by the negative bars, and it is being exported at Molinari. Total nitrogen concentrations were under the detection limit at Lebanik.

Figure 42

November Change in Nutrient Load per Linear Foot

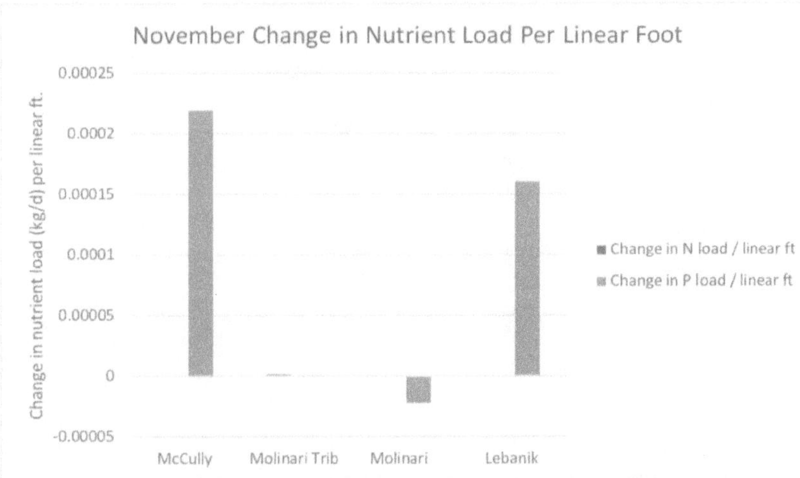

Note: Change in nutrient load per linear foot in November 2020 was different than the

July results. Total nitrogen data were limited in November due to undetectable amounts

of nitrogen in the samples. A small amount of nitrogen was exported at Molinari

Tributary. Phosphorus was exported at all sites except for Molinari where it was retained.

Figure 43

March Change in Nutrient Load per Linear Foot

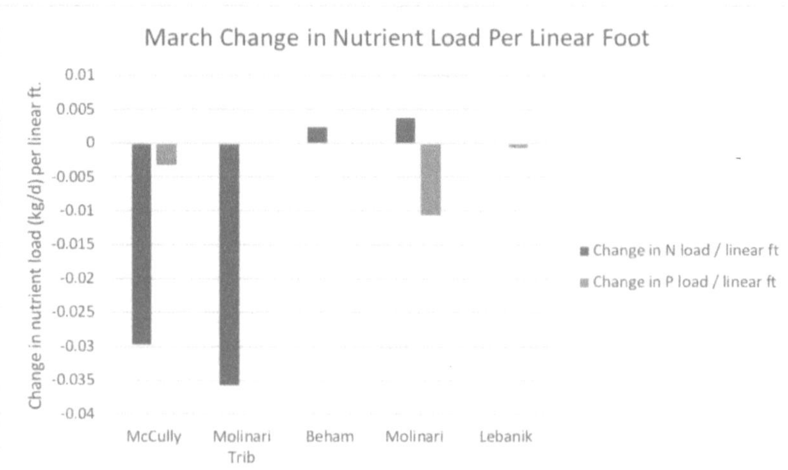

Note: Change in total nitrogen and total phosphorus load differed in March 2021, likely because of the high flow rate. Nitrogen was retained at McCully and Molinari Tributary and exported at Beham and Molinari. There was no change at Lebanik. Phosphorus was retained at McCully, Molinari, and Lebanik. There was no change at Molinari Tributary and Beham.

3.3.3. Nutrient Interactions Within the Floodplain

Tests were conducted to analyze interactions between sediment, surface water, and pore water nutrients in the floodplain wetlands. The data suggested that the correlation between surface water total nitrogen and sediment total nitrogen was not significant ($r = 0.34$, $N = 22$, $p = 0.12$). For total phosphorus, there was also no correlation between surface water and sediment concentrations ($r = 0.08$, $p = 0.73$). Surface water and sediment nitrogen concentrations were plotted, separated by restoration status, to further visualize the trends (Figure 44). The same was done for phosphorus concentrations (Figure 45). At the post-restoration sites, as surface water total nitrogen increases, sediment total nitrogen tends to as well. At pre-restoration sites, the opposite seems to be true. Total nitrogen concentrations were higher in both at the post-restoration sites. No trend can be seen between total phosphorus concentrations in the surface water and sediment, but concentrations were also higher for both at post-restoration sites.

Figure 44

Surface Water and Sediment Total N Concentrations

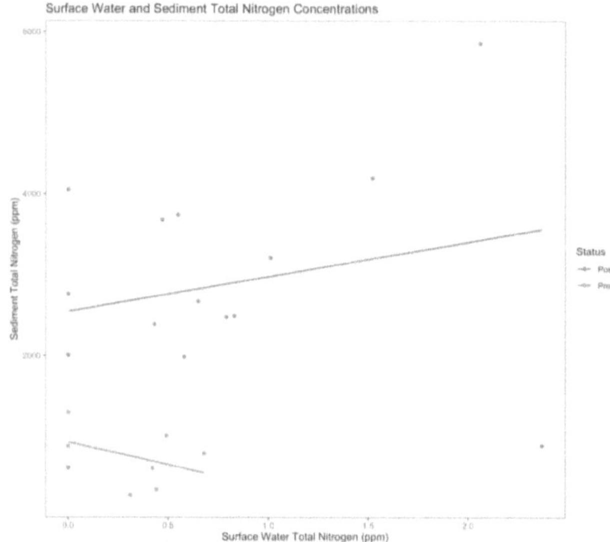

Note: The relationship between surface water total nitrogen and sediment total nitrogen was different when compared by restoration status. Surface water total nitrogen concentrations were variable, but sediment total nitrogen concentrations were generally higher at restored sites.

Figure 45

Surface Water and Sediment Total P Concentrations

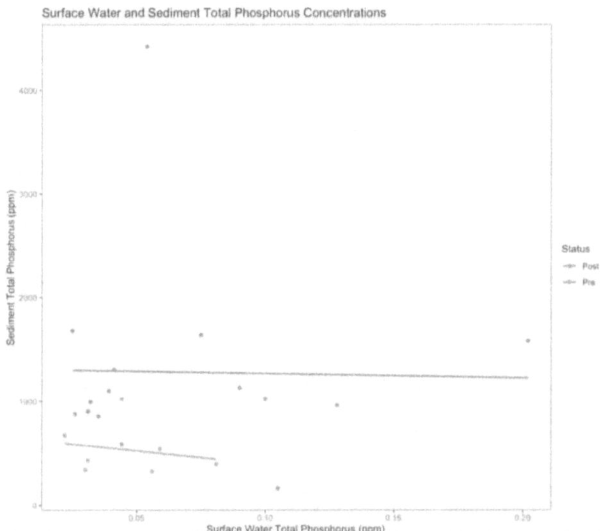

Note: Surface water total phosphorus and sediment total phosphorus were compared by restoration status. Post-restoration sediment total phosphorus concentrations were generally higher than pre-restoration.

Pore water and surface water nutrient concentrations were compared. Balance dissolved nitrogen is the difference between dissolved nitrogen and the sum of dissolved nitrate/nitrite and dissolved ammonia, and balance dissolved phosphorus is the difference between dissolved phosphorus and dissolved ortho phosphorus. The balance measurement represents the organic portion of the sample. This was calculated so that forms of pore water and surface water nutrients could be directly compared. Nitrogen makeup was predominately dissolved nitrate and nitrite as well as dissolved ammonia for both pore water and surface water. Dissolved ortho phosphorus dominated the dissolved phosphorus concentration in the surface water, and dissolved ortho phosphorus and balance dissolved phosphorus makeup was similar in the pore water. There was no correlation between balance dissolved nitrogen in the pore water and the surface water ($r = -0.27$, $N = 13$, $p = 0.38$). There was also no correlation between balance dissolved phosphorus in the pore water and surface water ($r = 0.005$, $p = 0.99$). This finding suggests that pore water and surface water nutrient concentrations were independent of one another.

A t-test was used to determine if there was a difference between surface water and pore water nutrient concentrations. There was no difference between balance dissolved nitrogen concentrations in the surface water versus pore water ($W = 66$, $p = 0.66$) (Figure 46). However, significantly more balance dissolved phosphorus was detected in the pore water versus the surface water ($W = 104$, $p = 0.017$) (Figure 47). This result suggests that the pore water stored more phosphorus than the surface water, but no conclusions could be drawn for nitrogen.

Figure 46

N Concentrations by Sampling Event

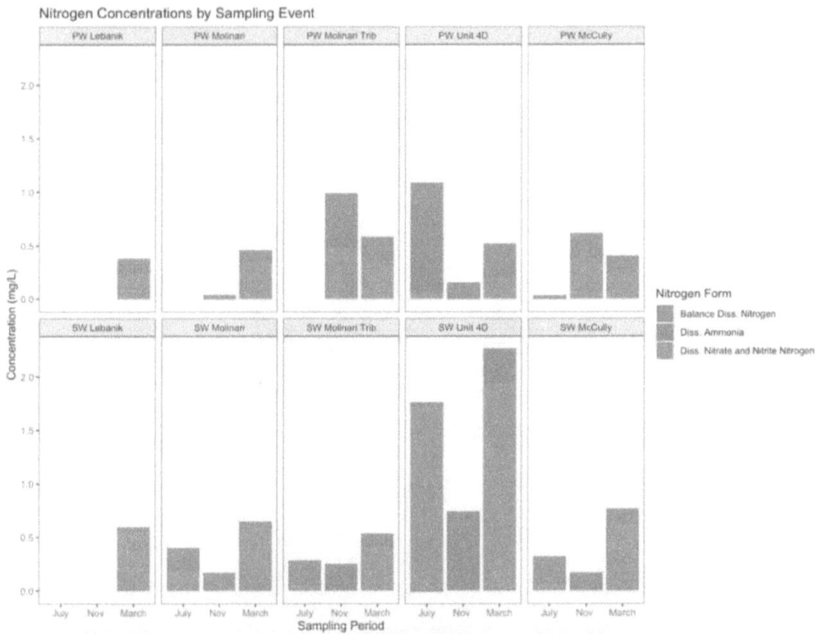

Note: Pore water dissolved nitrogen concentrations are shown on the top row, and downstream surface water dissolved nitrogen concentrations are shown on the bottom. There is a bar for each month that sampling occurred, with July and November happening in 2020 and March in 2021. Concentrations for both appeared to be dominated by dissolved nitrate and nitrite and balance dissolved nitrogen.

Figure 47

P Concentrations by Sampling Event

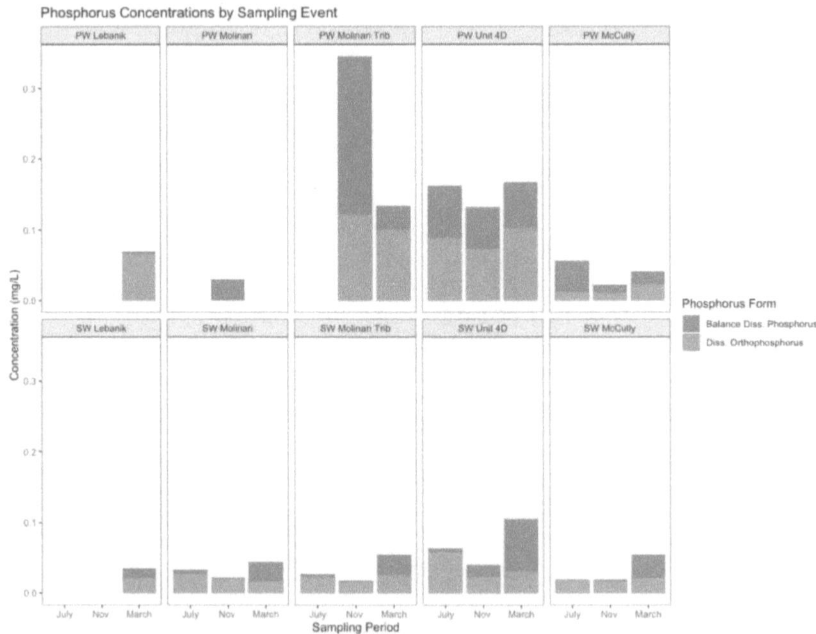

Note: Forms of dissolved phosphorus are shown, separated by pore water values on top and surface water on the bottom then further divided into sampling date. Dissolved balance phosphorus concentrations were greater in the pore water (W = 104, p = 0.017). Dissolved ortho phosphorus appeared to dominate the samples.

Chapter 4: Discussion

Data collection and analysis revealed several differences between restored and unrestored streams. Flow regime differed during these three sampling events, something that had an effect on sediment and nutrient dynamics across sites. Sediment grain size distribution and nutrient concentrations differed by restoration status; on average, coarse sediment was more common in the unrestored than restored sites and fine sediment was more common in the restored sites. Downstream sediment loading, retention, and export seemed to be more closely related to changes in flow rather than restoration status. Sediment nutrient concentrations, however, were found to be higher in the restored sites for both nitrogen and phosphorus. Trends in surface water nutrient data showed that during high flow, nutrient concentrations as well as loading may be decreased. The dissolved N:P ratio was higher at restored sites, but there was no difference for the total or suspended N:P ratio. Interactions between floodplain nutrients were also studied. Nitrogen and phosphorus trends were typically similar throughout, but only balance dissolved phosphorus was found to be higher in the pore water compared to the surface water. These associations cannot clearly explain the impact that floodplain reconnection has on restored streams, but they point to factors of interest for future research. Many other watershed features play a part in water, sediment, and nutrient storage in aquatic ecosystems; they should all be considered when identifying a potential restoration method.

4.1 Altered Flow Regime and Water Storage Capability

Hydrographs from the restored sites depicted changes in water depth before and after restoration. The range in water depth was typically smaller post-restoration, but the opposite was true at Beham. Less variation in water depth over time suggests that water generally spread laterally to the floodplain at restored sites, something that is beneficial for flood control and was an original goal of this restoration method (Wohl et al., 2015). In an in-depth look at responses to precipitation data at these study sites, Pazol (2021) found that the connected floodplains are being saturated post-restoration. This tends to increase transient storage and decrease peak flow, even during precipitation events. The change in hydrology at the restored sites is due to the removal of legacy sediment and the resulting floodplain connectivity.

Increased subsurface exchange as a means of heightened transient water storage is a common goal of restoration. Increased water storage can benefit water quality by improving nutrient retention and toxin removal mechanisms (Rana et al., 2017, Bukaveckas, 2007). Another study assessed transient storage as a means of conceptualizing nutrient uptake in a stream, something that highlights a stream's ability to store nutrients (O'Connor et al., 2009). Our study sites did not exhibit changes in water storage based on sampling event. However in November and March, primary headwater streams had higher flow rates than headwater streams did on average. This suggests that the smallest sites could hold more flow than the headwater sites during base and high flow despite having only 10% of the drainage area. Both pre- and post-restoration sites stored water more often than not, and no significant difference could be detected between the presence or absence of storage by date, so it is unclear what impact restoration had on the capability of these streams to store water. If more frequent sampling occurs, then long-term water storage could be investigated. Transient storage was expected to be higher in restored streams. In a study that used conservative salt tracer injections to conceptualize residence time, Bukaveckas (2007) found that transient storage was significantly higher at sites restored through floodplain reconnection. The author suggests that sediment makeup can limit hyporheic exchange. Fine grain clay sediment is commonly found in mid-Atlantic region streams like our sample sites in Pennsylvania (Merritts et al., 2007), so this could lessen the capacity of transient storage. Further, flow data was not consistently gathered over the three sampling events in this study due to various field limitations, so the overall water storage picture described here is incomplete. Difficulty reading flow due to low water, high water, and issues with the YSI conductivity meter led to this gap in water storage data. Bukaveckas (2007) also injected nitrogen and phosphorus into the stream with the salt addition in his study and found that reduced water velocity at restored sites was the leading cause for nutrient uptake enhancement. This allows for direct conceptualization of nutrient transport in streams, something that could benefit this study. Continuing to quantify transient water storage may be of future research interest at these study sites.

4.2 Effects on Sediment Retention and Export

Sediment dynamics differed by restoration status and flow regime. On average, unrestored sites had a higher proportion of coarse sediment deposited downstream than restored sites did, although the difference was not significant. Restored sites had a higher proportion of fine sediment on average than the unrestored sites, but this relationship also was not significant. Though smaller grain sizes are typically found in areas of slower water velocity (Farenhorst & Bryan, 1995)—an intended goal of restoration—no significant difference in flow was found between the restored and unrestored sites, so restoration status was likely not a direct cause of grain size makeup. Direct analyses between flow and sediment grain size found that coarse and medium sediment grain sizes were distributed independently of flow, but the proportion of fine grain sediment increased as flow did. While correlations between flow and coarse and medium sediment were not significant, there appeared to be a negative relationship for both; as flow increased, the proportion of medium and coarse sediment appeared to decrease. These relationships between flow and fine, medium, and coarse grain distribution are contrary to the idea that higher water velocity has the ability to transport larger sediment grain sizes (Beschta & Platts, 1986). However, lowered water velocity can still accompany high flow rates in the restored sites due to the floodplain's ability to receive water. Water velocity should be measured to investigate its effects on grain size distribution. In these study sites, flow seems to drive higher amounts of fine sediment and lower amounts of coarse and medium sediment. However, changes in experimental design could further clarify these interactions. In this study, the sediment samples are a composite of the sediment that was deposited into the traps in between sampling events. This means that comparing flow, a measurement at one moment in time, to sediment grain size distribution may not represent the whole relationship. If more frequent flow measurements were taken, the relationship between sediment grain size deposition and flow could be plotted as a continuum. While there were several instances where the sediment traps could not be accessed (Kent, Dunkard Fork, Molinari, Beham, and Lebanik in July 2020, Dunkard Fork in March 2020), samples from restored and unrestored sites were equally represented in the data. In several cases in November and March, a trowel was used to collect a sediment sample from the research sites. This could be a possible source of bias in the grain size data, and future efforts to standardize sampling by compositing a sample from several points in the stream should be made.

The movement of sediment within the channel was investigated through downstream TSS loading, sediment deposition, and overall export. Downstream TSS loading was variable by site and sampling event. There was no difference based on restoration status but increases in downstream TSS load were positively associated with high flow rates, a relationship that has been identified in previous research (Borris et al., 2013). There was also no difference between TSS load export per square mile by restoration status, likely because flow rate did not differ significantly by restoration status in this study. TSS is a complex variable because there are many elements that determine its movements at a watershed scale. The ability of restored waterways to capture and store solids can depend on the amount of TSS load coming from upstream, the morphological design of the restoration project, and weather characteristics like storm frequency and severity (Filoso et al., 2015). While this study identifies flow as a driver of downstream sediment transport, other watershed level characteristics should be considered to characterize differences in sediment movement based on channel morphology.

Sediment retention is a goal of floodplain reconnection restoration because of its influence on downstream water quality (Wohl et al., 2015). Sediment retention is identified by a higher TSS load upstream than what is found downstream, and this change in TSS load is the sediment deposition rate. There was no significant difference in sediment deposition at restored sites by sampling event, but sediment deposition was near 0 in July, positive in November, and negative in March at all sites, meaning that sediment retention was happening at high flow in March, but not at low flow in November. Sediment deposition per linear foot was low across all sampling events, ranging from -0.02 to 2.0 in the restored sites. Based on the presence of this relationship in other studies, it was expected that lower flow would encourage sediment deposition (Fennessy et al., 1994). It is known that water velocities are generally lower in restored reaches and higher in the unrestored reaches (McMillan & Noe, 2017), so it is possible that deposition could have been taking place on the floodplains of the restored reaches during and after high flow events. This was seen in a study by Ahilan et al. (2016). The authors found that in a restored floodplain, 20-30% of upstream sediment was deposited on the floodplain (Ahilan et al., 2016). Floodplain vegetation is known to enhance sediment deposition under periods of high flow (Abt et al., 1994), so this may explain the lack of in-stream sediment deposition observed in the restored sites in this study. With these findings in mind, it would be helpful to design an experiment that could quantify the amount of sediment deposited in the floodplain rather than in the channel. Once more data is collected during differing hydrologic conditions, the relationship between flow and sediment deposition at the restored sites may become clearer. Water velocity data should also be examined so that the difference between water velocity between restored and unrestored sites can be identified, and this can be compared with the presence of sediment deposition.

4.3 Factors Influencing Nutrient Cycling

4.3.1 Sediment Nutrients

The data showed that there was more sediment total nitrogen and phosphorus in the restored sites than the unrestored, suggesting that the sites are behaving as sediment nutrient sinks. This is an intended feature of floodplain reconnection; sediment concentrations of nitrogen and phosphorus have been shown to increase at restored streams which can improve water quality further downstream (Noe et al., 2019). Sediment adsorption of nutrients is a form of temporary storage; smaller grain sizes are more likely to adsorb nutrients, so it follows that the restored sites retained nutrient-rich fine-grained sediment (Newcomer Johnson et al., 2016). Sediment nutrient concentrations were not associated with variations in flow in this study, though others have found that the presence of sediment nitrogen and phosphorus increased with flow rate in floodplains (Dalu et al., 2019). When periods of high flow occur in the restored sites and the channel connects with the floodplain, it is likely that floodplain soil will interact with the channel sediment. This floodplain soil is usually rich in nutrients due to the ability of wetland vegetation to capture and breakdown nutrients at a high rate (Madsen & Cedergreen, 2002), so this likely has an influence on the nutrient composition of in-stream sediment.

4.3.2 Surface Water Nutrients

Surface water nutrient concentrations were variable across restoration status and sampling date. Flow was not statistically related to total, dissolved, or suspended surface water nutrients. However, visualizations of the data suggest that higher flow rates were associated with lower concentrations of total nitrogen and phosphorus. Pennsylvania does not currently have water quality criteria standards for streams. One report (Capacasa, 2008) created a watershed-specific TMDL for phosphorus through stressor-response analyses and a culmination of values from the literature. Researchers found that at Indian Creek in Montgomery County, Pennsylvania (roughly 300 miles from the study sites in our research), a stream with a drainage area of 7 square miles, a concentration of 0.04 mg/L total phosphorus is recommended daily from April 1-October 31. Our study sites had average total phosphorus concentrations slightly above this limit in July for both unrestored sites (0.062 mg/L) and restored sites (0.084 mg/L). However, the TMDL for Indian Creek also factors in the need to limit phosphorus input due to excessive amounts from nearby residences and agriculture. Since there are no state standards, it is difficult to gauge how our nutrient concentrations compare.

N:P ratios are often used to assess the makeup of nutrients in an aquatic ecosystem to determine the presence of limiting nutrients. The EPA reports that if the N:P ratio is < 10, nitrogen is the limiting nutrient. If N:P is > 10, phosphorus is the limiting nutrient (EPA, 2000). In this dataset, total N:P ranged from 0-22, dissolved ranged from 0-48, and suspended ranged from 0-15. The majority of N:P in all forms was < 10, suggesting that nitrogen is the limiting nutrient in these ecosystems. Compared to

13 streams with varying surrounding land use across the United States in Manning et al. (2020), our N:P ratios are much lower. No significant difference between total and suspended N:P ratios at restored versus unrestored sites was identified in this study. However, dissolved N:P ratio was higher on average in restored sites. These ratios were not statistically correlated with flow, but it appears that as flow increased, dissolved N:P might increase slowly. Particulate surface water nutrients are known to increase with flow (Stutter et al., 2008), but the influence of flow on dissolved nutrients may be less direct. Stream water N:P ratios vary based on climate, sediment makeup, watershed storage, and human influence (Green et al., 2007). N:P ratios should be compared with biological diversity and abundance data so that the impact of surface water nutrient ratios on biology can be identified. Stelzer and Lamberti (2001) found that the abundance of nine out of eleven algal taxa was affected by the N:P ratio as well as nutrient concentration. Comparing biological data from our study sites with N:P ratio and nutrient concentration data can identify the magnitude of the effect that surface water nutrients have on biology.

Surface water nitrogen and phosphorus loading quantifies the amount of nutrients moving through a point over time, something that is helpful for understanding the extent of nutrient retention in a stream. Downstream nutrient loading, a measure of nutrient load export, was not different between restored and unrestored sites or between stream size classes. Both total nitrogen and total phosphorus downstream loading was found to be driven by flow, something that builds on the evidence suggesting the influence that flow has on nutrient presence in the study sites. At the restored sites where upstream and downstream data was available, a higher nitrogen load was seen downstream compared to upstream at every site in July during the low flow period. During high flow in March, a higher nitrogen load was observed upstream at every site but Molinari (where the load was similar upstream and downstream). This suggests that the sites are retaining nitrogen in high flow and exporting nitrogen in low flow. This may be due to the stream's ability to expand laterally and reduce water velocity under high flow conditions. Phosphorus loading was variable. Dissolved nutrient load typically outweighed suspended load for both nitrogen and phosphorus during low flow conditions in July and November 2020. In March 2021, during a period of high flow, dissolved and suspended phosphorus loads were similar, but nitrogen loads were still dominated by dissolved rather than suspended. This indicates that high flow may be related to an increase in suspended phosphorus loading, something that has been observed in the literature (Farenhorst & Bryan, 1995, Corriveau et al., 2013).

Nutrient load export per square mile data was not statistically different between restored and unrestored sites, but it appears that unrestored sites generally had lower export per square mile. The largest streams generally had lower amounts of export per square mile compared to the smaller streams. Nutrient load retention was variable. In the restored sites, phosphorus load was retained at two out of the three sites in July and nitrogen load was exported at all. In November, nitrogen concentrations were below detection levels at all sites except for Molinari Tributary, Unit 4D, and McNay—all sites with drainage area less than 1 square mile—but phosphorus load was retained at one out of four sites. In March, nitrogen load was retained at two sites, exported at two sites, and no change was detected at one site. Phosphorus load was retained at three sites and no change was detected at two. This suggests that high flow might encourage higher nitrogen load export per square mile, something that has been noted in the literature (Boynton & Kemp, 2000). No relationship is seen with phosphorus, though this has also been observed in the literature (Pionke et al., 1996). These results together suggest that nutrient load export data is too variable to make any conclusions, so more data are needed to determine drivers of nutrient export. The mechanisms behind nitrogen transport are different than those that drive phosphorus, so they should be considered more independently in the future.

Change in nitrogen loading from upstream to downstream was used as a benchmark for denitrification rate in restored reaches where a smaller nitrogen load downstream indicated that denitrification occurred along the reach. In July, the difference in nitrogen load per linear foot from upstream to downstream ranged from 0-0.04 kg/d. In November, when nitrogen concentrations were low, the difference ranged from 0-0.000001 kg/d which is an insignificant amount. This suggests that in low flow conditions, a low nitrogen input is observed at the restored sites. There was more nitrogen downstream than upstream at every site in July and November, so there was some mobilization of nitrogen within the stream rather than denitrification at low flow. In March, the difference ranged from -0.04-0.004 kg/d. Denitrification occurred at Molinari Tributary and McCully. These sites had flow rates near the median. This suggests that denitrification was not happening at a high rate during low or high flow conditions. The low nitrogen input during low and base flow was a main contributing factor to low denitrification rates in July and November. A study by McMahon et al. (2021) found that the main drivers of increased nitrogen loading were precipitation and land use rather than restoration via floodplain reconnection. Their results showed that denitrification rates were relatively unchanged 1-5 years after restoration (McMahon et al., 2021). The authors of that study suggest that carbon availability limits denitrification, so future research should include an analysis on surface water carbon. After several more years of sampling, this relationship should be revisited to see under what flow conditions denitrification is likely to occur, or if the relationship changes as the restored reaches become more mature.

4.3.3. Floodplain Nutrient Interactions

No significant correlations were identified between surface water and sediment total nutrient concentrations. Other studies have found that if surface water nutrient concentrations are high, sediment particles will adsorb those nutrients (Vought et al., 1994). It appears that there was a slight positive trend between the sediment and surface water data at post-restoration sites and a slight negative trend at pre-restoration sites. There appears to be no relationship between surface water and sediment phosphorus concentrations at restored and unrestored sites. These results suggest that a positive relationship between surface water and sediment total nitrogen might be present at post-restoration sites, implying that the two might interact more than they do at pre-restoration sites. The relationship between sediment and surface water nutrient concentrations is not well understood and would benefit from further research. It is also unclear whether biota benefits more from sediment or surface water nutrient concentrations, so this should also be investigated in future studies.

There was no correlation between surface water and pore water nutrients. This could be due to the lack of pore water data that was available in this study. Pore water data were only collected at five of the restored sites, and it was often difficult to collect enough to fill the sample volume requirements at low flow conditions. Other research suggests that surface water and pore water commonly interact in the hyporheic zone, and stream morphology and transient flow are drivers of these interactions (Hillel et al., 2019). There are likely many different factors that control these two systems, so they should be considered separately when identifying pathways of nutrient cycling. A larger dataset is needed to determine how nutrients are transferred through these three media.

More organic phosphorus was found in the pore water compared to the surface water at restored sites. No significant difference was observed for nitrogen concentrations, but surface water concentrations were generally higher. Wetland vegetation is known to catch and foster the breakdown of nutrients before they enter the surface water, something that might explain the increase in phosphorus concentrations in the pore water; this phenomenon is commonly seen with nitrogen also (Vought et al., 1994). The discrepancy between nitrogen and phosphorus trends can be explained by the low concentrations of nitrogen in combination with unavailable data. The lab was unable to detect nitrogen concentrations from several of the pore water and surface water nitrogen samples due to low sample volume and in one case, the sample holding time was exceeded. Beyond sampling issues, the nitrogen inputs were low under low flow conditions. This made it difficult to assess any relationship between pore water and surface water concentrations. More variations in the data are needed to identify drivers of pore water and surface water nutrient interactions.

Chapter 5: Conclusion

This study provides preliminary research on the impact of floodplain reconnection as a restoration method in southwest Pennsylvania. It was hypothesized that the restored streams would retain water, sediment, and nutrients compared to unrestored streams. Water storage, sediment dynamics, and nutrient cycling were analyzed to understand how post-restoration streams behave compared to those that are unrestored. There was no difference between short-term water storage at restored versus unrestored sites, but long-term storage should be investigated if an increase in sampling frequency can be achieved. Sediment composition and transport were conceptualized. Sediment grain size distribution differed by restoration status; unrestored streams were comprised of more coarse sediment, and fine sediment was more abundant in restored streams. Flow was positively related with fine grain sediment abundance, but it seemed to have a negative relationship with coarse and medium sediment. While no significant difference was discovered between sediment deposition based on restoration status, it appears as though sediment deposition was most likely to occur during low flow while resuspension was likely during high flow. Downstream TSS loading and overall TSS retention were not affected by restoration status, but both appear to be driven by high flow events. This evidence signifies that the rate of sediment retention might be more closely related to flow, something that was not found to be different based on restoration status during these sampling events. An investigation into nutrient cycling suggested that sediment bound nutrients were more abundant at restored streams, a relationship that is likely due to the increased presence of fine grain sediment at restored sites. The data also suggested that high flow rates led to low surface water total, dissolved, and suspended nitrogen and phosphorus concentrations at the restored sites. Other analyses showed that when flow rate was low, as it was in the two 2020 sampling events, a low nitrogen input is observed at the restored sites. Surface water nitrogen loading amounts were higher upstream during high flow and lower during low flow, suggesting that flow is a driver of upstream nitrogen input. Surface water nutrient retention was variable across restoration status, but it appeared to be greatest under high flow conditions. Again, denitrification did not seem to be associated with restoration status. Rather, it was more likely to happen during median flow rates. Surface water nitrogen was mostly comprised of dissolved nitrate and nitrite nitrogen while pore water nitrogen forms were more variable. Balance dissolved phosphorus was higher in the pore water than it was in the surface water at restored sites, something that provides insight to organic phosphorus makeup across the floodplain.

These data are beginning to answer important questions regarding the effectiveness of floodplain reconnection restoration. When choosing a stream restoration method, it is important to think holistically about the watershed. Land use, storm events, and legacy factors should be considered before deciding on a method. These physical factors should be studied in conjunction with the growing dataset behind floodplain reconnection to create a watershed-level restoration plan. These data can be improved upon with more consistent sampling efforts. This study has several instances where data could not be collected, so additional information is needed before making definite conclusions. However, the data in this study serve as a beginning assessment of long-term work that will be completed on these study sites. A more abundant dataset will lead to a greater understanding of the restored sites' reactions to variations in hydrology, nutrient loading, and the effect that time and establishment has on ecosystem quality. This research serves as an addition to the relatively young body of work on the floodplain reconnection method of restoration.

Future work should focus on further identifying mechanisms of water storage through long-term data collection. Quantities of floodplain sediment deposition should be studied to assess how sediment is transported in restored systems. This will help clarify if sediment is actually exported downstream or if it is deposited onto the floodplain. Further, water velocity should be investigated so that it can be directly compared with deposition rates. It would also be helpful to study these sites long-term so the impact of ecosystem establishment can be assessed. Denitrification rate, water storage, and nutrient retention might change as the restored floodplains grow to be more cohesive.

References

Abt, S. R., Clary, W. P., Thornton, C. I. (1994). Sediment deposition and entrapment in vegetated streambeds. *Journal of Irrigation and Drainage Engineering, 120*(6), 1098-1111.

Ahilan, S., Guan, M., Sleigh, A., Wright, N., Chang, H. (2016). The influence of floodplain restoration on flow and sediment dynamics in an urban river. *Flood Risk Management, 11*(S2), S986-S1001.

Azinheira, D. L., Scott, D. T., Hession, W., Hester, E.T. (2014). Comparison of effects of inset floodplains and hyporheic exchange induced by in-stream structures on solute retention. *Water Resource Research, 50,* 6168-6190.

Bayley, P. B. (1991). The flood pulse advantage and the restoration of river-floodplain systems. *Regulated Rivers: Research and Management, 6,* 75-86.

Beschta, R. L., Platts, W. S. (1986). Morphological features of small streams: significance and function. *Journal of the American Water Resources Association, 22*(3), 369-379.

Borris, M., Viklander, M., Gustafsson, A-M., Marsalek, J. (2013). Modelling the effects of changes in rainfall event characteristics on TSS loads in urban runoff. *Hydrological Processes, 28*(4), 1787-1796.

Boynton, W. R., Kemp, W. M. (2000). Influence of river flow and nutrient loads on selected ecosystem processes. *Estuarine Science,* 269-298.

Bukaveckas, P. A. (2007). Effects of channel restoration on water velocity, transient storage, and nutrient uptake in a channelized stream. *Environmental Science & Technology, 41,* 1570-1576.

Capacasa, J. (2008). *Nutrient and sediment TMDLs for the Indian Creek Watershed, Pennsylvania: Established by the U.S. Environmental Protection Agency.* U.S. Environmental Protection Agency. https://ofmpub.epa.gov/waters10/attains_impaired_waters.show_tmdl_document? p_tmdl_doc_blobs_id=75058

Cedergreen, N., Madsen, T. V. (2002). Sources of nutrients to rooted submerged macrophytes growing in a nutrient-rich stream. *Freshwater Biology, 47*(2), 283-291.

Corriveau, J., Chambers, P. A., Culp, J. M. (2013). Seasonal variation in nutrient export along streams in the Northern Great Plains. *Water, Air, and Soil Pollution, 224.*

Craig, L.S., Palmer, M., Filoso, S., Bernhardt, E. S. (2008). Stream restoration strategies for reducing river nitrogen loads. *Frontiers in Ecology and the Environment, 6*(10), 529-538.

Dalu, T., Wasserman, R. J., Magoro, M. L., Froneman, P. W., Weyl, O. L. F. (2019). River nutrient water and sediment measurements inform on nutrient retention with implications for eutrophication. *Science of the Total Environment, 684,* 296-302.

Davis, R. T., Tank, J. L., Mahl, U. H., Winikoff, S. G., Roley, S. S. (2015). The influence of two stage ditches with constructed floodplains on water column nutrients and sediments in agricultural streams. *Journal of the American Water Resources Association, 51,* 941-955.

Ding, S., Chen, M., Gong, M., Fan, X., Qin, B., Xu, H., Gao, S., Jin, Z., Tsang, D. C. W., Zhang, C., (2018). Internal phosphorus loading from sediments causes seasonal nitrogen limitation for harmful algal blooms. *Science of the Total Environment, 625,* 872-884.

Domagalski, J. L., Johnson, H. (2012). Phosphorus and groundwater: Establishing links between agricultural use and transport to streams. Retrieved from: https://pubs.usgs.gov/fs/2012/3004/

EPA. (2000). Nutrient criteria technical guidance manual. Retrieved from: https://www.epa.gov/sites/production/files/2018-10/documents/nutrient-criteria-manual-rivers-streams.pdf

Farenhorst, A., Bryan, R. B. (1995). Particle size distribution of sediment transported by shallow flow. *CATENA 25*(1-4), 47-62.

Fennessy, M. S., Brueske, C. C., Mitsch, W. J. (1994). Sediment deposition patterns in restored freshwater wetlands using sediment traps. *Ecological Engineering, 3,* 409-428.

Filoso, S., Smith, S. M., Williams, M. R., Palmer, M. A. (2015). The efficacy of constructed stream-wetland complexes at reducing the flux of suspended solids to Chesapeake Bay. *Environmental Science & Technology, 49,* 8986-8994.

Gowan, C., Fausch, K. D. (1996). Long-term demographic responses of trout populations

to habitat manipulation in six Colorado streams. *Ecological Applications, 6,* 931-

946.

Green, M. B., Johnson, G., Magner, J., Schaefer, B. (2007). Flow path influence on an

N:P ratio in two headwater streams: A paired watershed study. *Journal of

Geophysical Research, 112*(G3), 1-11.

Harrison, M. D., Miller, A. J., Groffman, P. M., Mayer, P. M., Kaushal, S. S. (2014).

Hydrologic controls on nitrogen and phosphorus dynamics in relict oxbow

wetlands adjacent to an urban restored stream. *Journal of the American Water

Resources Association, 50,* 1365-1382.

Hartranft, J. (2019). Big Spring Run restoration project background & monitoring results

[Video webinar]. Retrieved from:

https://chesapeakestormwater.net/events/big_spring_run_research/

Hefting, M., Clément, J. C., Dowrick, D., Cosandey, A. C., Bernal, S., Cimpian, C.,

Tatur, A., Burt, T. P., Pinay, G. (2003). Water table elevation controls on soil

nitrogen cycling in riparian wetlands along a European climatic gradient.

Biogeochemistry, 67, 113-134.

Hester, E. T., Young, K. I., Widdowson, M. A. (2013). Mixing of surface and

groundwater induced by riverbed dunes: implications for hyporheic zone

definitions and pollutant reactions. *Water Resources, 49*(9), 5221-5237.

Hillel, N., Wine, M. L., Laronne, J. B., Licha, T., Be'eri-Shelvin, Y., Siebert, C. (2019).

Identifying spatiotemporal variations in groundwater-surface water interactions using shallow pore water chemistry in the lower Jordan river. *Advances in Water Resources 131,* 103388-103398.

Hupp, C. R., Noe, G. B., Schenk, E. R., Benthem, A. J. (2013). Recent and historic sediment dynamics along Difficult Run, a suburban Virginia Piedmont stream. *Geomorphology, 180,* 256-269.

Junk, W. J., Bayley, W. P., Sparks, R. E. (1989). The flood pulse concept in river-floodplain systems. *Aquatic Science, 106,* 110-127.

Kaushal, S. S., Groffman, P. M., Mayer, P. M., Striz, E., Gold, A. J. (2008). Effects of stream restoration on denitrification in an urbanizing watershed. *Ecological Applications, 18,* 789-804.

Kondolf, G. M. (1995). Geomorphological stream channel classification in aquatic habitat restoration: Uses and limitations. *Aquatic Conservation, 5*(2), 127-141.

Larson, L. N., Kipp, G. G., Mott, H. V., Stone, J. J. (2012). Sediment pore-water interactions associated with arsenic and uranium transport from the North Cave Hills mining region, South Dakota, USA. *Applied Geochemistry, 27,* 879-891.

Manning, D. W. P., Rosemond, A. D., Benstead, J. P., Bumpers, P. M., Kominoski, J. S. (2020). Transport of N and P in U.S. streams and rivers differs with land use and between dissolved and particulate forms. *Ecological Applications, 30*(6), 1-17.

MARC. What is sediment pollution? No date. Retrieved from: https://cfpub.epa.gov/npstbx/files/ksmo_sediment.pdf

McMahon, P., Beauchamp, V. B., Casey, R. E., Salice, C. J., Bucher, K., Marsh, M., Moore, J. (2021). Effects of stream restoration by legacy sediment removal and floodplain reconnection on water quality. *Environmental Research Letters, 16,* 1-13.

McMillan, S. K., Noe, G. B. (2017). Increasing floodplain connectivity through urban stream restoration increases nutrient and sediment retention. *Ecological Engineering, 108,* 284-295.

Merritts, D., Walter, R. C., Rahnis, M. (2010). *Sediment and nutrient loads from stream corridor erosion along breached millponds.* Franklin and Marshall College. https://files.dep.state.pa.us/Water/Chesapeake%20Bay%20Program/ChesapeakeP ortalFiles/Legacy%20Sediment%20Workgroup/Sediment%20and%20Nutrient%2 0Loads%20from%20Stream%20Corridor%20Erosion%20along%20Breached%2 0Millponds.pdf

Mitsch, W. J., Cronk, J. K., Xinyuan, W., and Nairm, R. M. (1995). Phosphorus retention in constructed freshwater riparian marshes. *Ecological Applications, 5,* 830–845.

Moore, R. D. (2003). Introduction to salt dilution gauging for streamflow measurement. *Watershed Management Bulletin, 7*(4), 20-23.

Newcomer Johnson, T. A., Kaushal, S. S., Mayer, P. M., Smith, R. M., Sivirichi, G. M. (2016). Nutrient retention in restored streams and rivers: a global review and synthesis. *Water, 8*(4), 116.

Noe, G. B. Hupp, C. R., Rybicki, N. R. (2013). Hydrogeomorphology influences soil nitrogen and phosphorus mineralization in floodplain wetlands. *Ecosystems, 16,* 75-94.

Noe, G. B., Boomer, K., Gillespie, J. L., Hupp, C. R., Martin-Alciati, M., Floro, K., Schenk, E. R., Jacobs, A., Stano, S. (2019). The effects of restored hydrologic connectivity on floodplain trapping vs. release of phosphorus, nitrogen, and ediment along the Pocomoke River, Maryland USA. *Ecological Engineering, 138,* 334-352.

NWS. No date. Retrieved from: http://climod2.nrcc.cornell.edu

O'Connor, B. L., Hondzo, M., Harvey, J. W. (2009). Predictive modeling of transient storage and nutrient uptake: implications for stream restoration. *Journal of Hydraulic Engineering, 136*(12), 1018-1032.

PA DCNR. No date. History of Ryerson Station State Park. Retrieved from: https://www.dcnr.pa.gov/StateParks/FindAPark/RyersonStationStatePark/Pages/History.aspx

Palmer, M.A., Bernhardt, E., Schlesinger, W., Eshleman, K., Foufoula-Georgiou, E. Hendryx, M., Lemly, A., Likens, G., Loucks, O., Power, M. (2010). Mountaintop mining consequences. *Science, 327*(5962), 148-159.

Palmer, M. A., Hondula, K. L., Koch, B. J. (2014). Ecological restoration of streams and rivers: Shifting strategies and shifting goals. *Annual Review of Ecology, Evolution, and Systematics, 45,* 247-269.

Pazol, J. (2021). *Effects of floodplain reconnection on storm response of restored river ecosystems.* [Unpublished undergraduate thesis]. Ohio University.

Pennsylvania Department of Environmental Protection. (2018). Water quality monitoring protocols for streams and rivers. Office of Water Programs.

https://files.dep.state.pa.us/Water/Drinking%20Water%20and%20Facility%20Re gulation/WaterQualityPortalFiles/Technical%20Documentation/MONITORING_ BOOK.pdf

Pionke, H. B., Gburek, W. J., Sharpley, A. N., Schnabel, R. R. (1996). Flow and nutrient export patterns for an agricultural hill-land watershed. *Water Resources Research, 32*(6). 1795-1804.

Poff, N. L., Olden, J. D., Merritt, D. M., Pepin, D. M. (2007). Homogenization of regional river dynamics by dams and global biodiversity implications. *PNAS, 104*(14), 5732-5737.

Polvi, L. E., Nilsson, C., Hasselquist, E. M. (2014). Potential and actual geomorphic complexity of restored headwater streams in northern Sweden. *Geomorphology, 210,* 98-118.

R Core Team. (2019). R: A language and environment for statistical computing. R Foundation for Statistical Computing, Vienna, Austria. https://www.R-project.org/

Rana, S. M. M., Scott, D. T., Hester, E. T. (2017). Effects of in-stream structures and channel flow rate variation on transient storage. *Journal of Hydrology, 548,* 157-169.

Reddy, K. R., Kadlec, R. H., Flaig, E., Gale, P. M. (1999). Phosphorus retention in streams and wetlands: A review. *Critical Reviews in Environmental Science and Technology, 29*(1), 83-146.

Roley, S. S., Tank, J. L., Stephen, M. L., Johnson, L. T., Beaulieu, J. J., Witter, J. D. (2012). Floodplain restoration enhances denitrification and reach-scale nitrogen removal in an agricultural stream. *Ecological Applications, 22*, 281-297.

Rosgen, D. L. (2007). Rosgen geomorphic channel design. National Engineering Handbook 654. United States Department of Agriculture, Washington, D.C. USA.

Schenk, E. R., Hupp, C. R. (2009). Legacy effects of colonial millponds on floodplain sedimentation, bank erosion, and channel morphology, Mid-Atlantic, USA. *Journal of the American Water Resources Association, 45*(3), 597-606.

Simon, A., Doyle, M., Kondolf, M., Shields, F. D. Jr., Rhoads, B., McPhillips, M. (2007). Critical evaluation of how the Rosgen classification and associated "natural channel design" methods fail to integrate and quantify fluvial processes and channel response. *Journal of the American Water Resources Association, 43*(5), 1117-1131.

SoilMoisture. No date. Agronomy. https://www.soilmoisture.com/home.php

Stelzer, R. S., Lamberti, G. A. (2001). Effects of N:P ratio and total nutrient concentration on stream periphyton community structure, biomass, and elemental composition. *Limnology and Oceanography, 46*(2), 356-367.

Stutter, M. I., Langan, S. J., Cooper, R. J. (2008). Spatial and temporal dynamics of stream water particulate and dissolved N, P, and C forms along a catchment transect, NE Scotland. *Journal of Hydrology, 350*(3-4), 187-202.

Surridge, Ben W. J., Heathwaite, A. L., Baird, Andy J. (2012). Phosphorus mobilization and transport within a long-restored floodplain wetland. *Ecological Engineering, 44,* 348-359.

Tockner, K., Pennetzdorfer, D. Reiner, N., Schiemer, F., Ward, J. V. (1999). Hydrological connectivity, and the exchange of organic matter and nutrients in a dynamic river floodplain system. *Freshwater Biology, 41,* 521-535.

U.S. Geological Survey. (2016). The StreamStats program. http://streamstats.usgs.gov

USGS. No date. Retrieved from: https://waterdata.usgs.gov/usa/nwis/uv?03112000

Vought, L. B.-M., Dahl, J., Pedersen, C. L., Lacoursière, J. O. (1994). Nutrient retention in riparian ecotones. *Ambio, 23*(6), 342-348.

Walter, R. C., Merritts, D. J. (2008). Natural streams and the legacy of water-powered mills. *Science, 319*(5861), 299-304.

Walter, R., Merritts, D., Rahnis, M., Langland, M., Galeone, D., Gellis, A., Hilgartner, W., Bowne, D., Wallace, J., Mayer, P., Forshay, K. (2013). Big Spring Run natural floodplain, stream, and riparian wetland—Aquatic resource restoration project monitoring. PA DEP Final Report.

Wohl, E., Lane, S. N., Wilcox, A. C. (2015). The science and practice of river restoration. *Water Resources Research, 51,* 5974-5997.

www.ingramcontent.com/pod-product-compliance
Lightning Source LLC
LaVergne TN
LVHW041714190726
843493LV00007B/2093